V-8 Horsepower
Performance Handbook

Scott Parkhurst

motorbooks

First published in 2009 by Motorbooks, an imprint of MBI Publishing Company, 400 First Avenue North, Suite 300, Minneapolis, MN 55401 USA

Motorbooks titles are also available at discounts in bulk quantity for industrial or sales-promotional use. For details write to Special Sales Manager at MBI Publishing Company, 400 First Avenue North, Suite 300, Minneapolis, MN 55401 USA.

To find out more about our books, visit us online at www.motorbooks.com.

Library of Congress Cataloging-in-Publication Data

Parkhurst, Scott, 1966–
    V-8 horsepower performance handbook / Scott Parkhurst.
      p. cm.
   ISBN 978-0-7603-3552-9 (sb : alk. paper)
   1. Automobile—Motors—Maintenance and repair. 2. Automobiles—Performance. 3. Automobiles—United States. I. Title.
   TL210.P327 2009
   629.25'04—dc22
                                    2009017865
ISBN-13: 978-0-7603-3552-9

Editor: Chris Endres
Designer: Jessica Foeller

Printed in China

About the Author
Scott Parkhurst's experience with high-performance V-8 engines began before he had his driver's license. Later, working as a technician in race-engine shops, he learned much about the professional preparation of high-output V-8s. Parkhurst then spent seven years as technical editor at *Popular Hot Rodding* magazine and co-founded the Engine Masters Challenge competition, developing its innovative format and writing its rules. The success of the Engine Masters Challenge led to *Engine Masters Quarterly* magazine, with Parkhurst as its founding editor. The opportunity to launch a new enthusiast magazine brought him to Minneapolis, Minnesota, where he was the founding editor of *Street Thunder*. After some three years guiding this performance publication, Parkhurst left *Street Thunder* to pursue other gearhead interests. He still works directly with enthusiasts to develop custom V-8s to suit their specific needs, and of course, in his spare time he builds engines for his own projects.

Front cover image
© Ron Kimball/Kimballstock

Contents

Introduction

From the birth of the mass production American V-8 in the 1930s to its rapid evolution during the horsepower wars of the 1960s and early 1970s, there was never any questioning the validity of the design. A relatively compact package capable of producing great power, the V-8 was needed to power the ever-larger cars coming out of Detroit, and it was up to the automakers' capable engineers to give buyers more and more power every year. They obliged, and the natural development of the great American V-8s ultimately led to the legendary muscle car era.

For hot rodders, these factory engines are just the beginning. Building a custom engine to suit the specific needs of a special project requires both knowledge and talent, and knowing where to begin is half the battle. The actual assembly of a typical American V-8 is not extremely difficult, but the design of the overall performance package requires a significant investment of research to get the most bang for your buck. Teaming a group of parts that function as an efficient system is well worth the time.

The purpose of this book is to assist you in this research. Some basic characteristics and measured values work in harmony to produce power, and whatever you'd like your engine of choice to accomplish, this book can help get you there. Some new technologies have been developed that can really help out older engines; these technologies and components are explored and explained here as well.

American V-8s will always be known for their power and reliability. After reading this book, you'll know why and how this is possible. You'll also know how to find even more power and durability, should you choose to build a custom engine of your own, regardless of make.

Chapter 1
History of the American V-8

THE FORD FLATHEAD

First produced in 1932, Ford's flathead engine was the first mass production American V-8. It was thoroughly researched prior to being offered, and it was this extensive development that can be credited with the quality of the final product. It was reliable, powerful, and relatively easy to work on. It was embraced by racers and hot rodders almost immediately. Its broad availability and reasonable price, and its power-per-dollar potential, made it the powerplant of choice for hot rodders from the 1930s to the 1950s. It wasn't until overhead valve designs became readily available and affordable that the flathead was dethroned. The Ford flathead is still popular today—not as a competition powerplant, but rather as a nostalgic reminder of the way things used to be.

In the late 1940s and early 1950s, the development and subsequent launch of overhead-valve V-8 designs by Cadillac and Oldsmobile revolutionized hot rodding. Moving the intake and exhaust valves out of the block and into the cylinder heads allowed a great increase in both quantity and quality of airflow to the combustion chambers. This design also allowed larger displacement, as the bore diameter was no longer limited by the need to clear the valves. Larger-displacement engines that could move more air and fuel in and out simply had greater power potential, and hot rodders were quick to realize this.

The major downside of the new overhead-valve design was its greater physical dimensions and added weight, especially when compared to the diminutive flathead. The heavier, bulkier new engines presented a challenge when the early rodders performed engine swaps, although the gains in power offset these disadvantages.

The Ford flathead was the first mass-production V-8, and was truly embraced by hot rodders. It's easily identified by its unique appearance, and while it came in many different forms throughout its production run from 1932 to 1954, the look didn't change much. This one is a just a model, but the appearance is distinct.

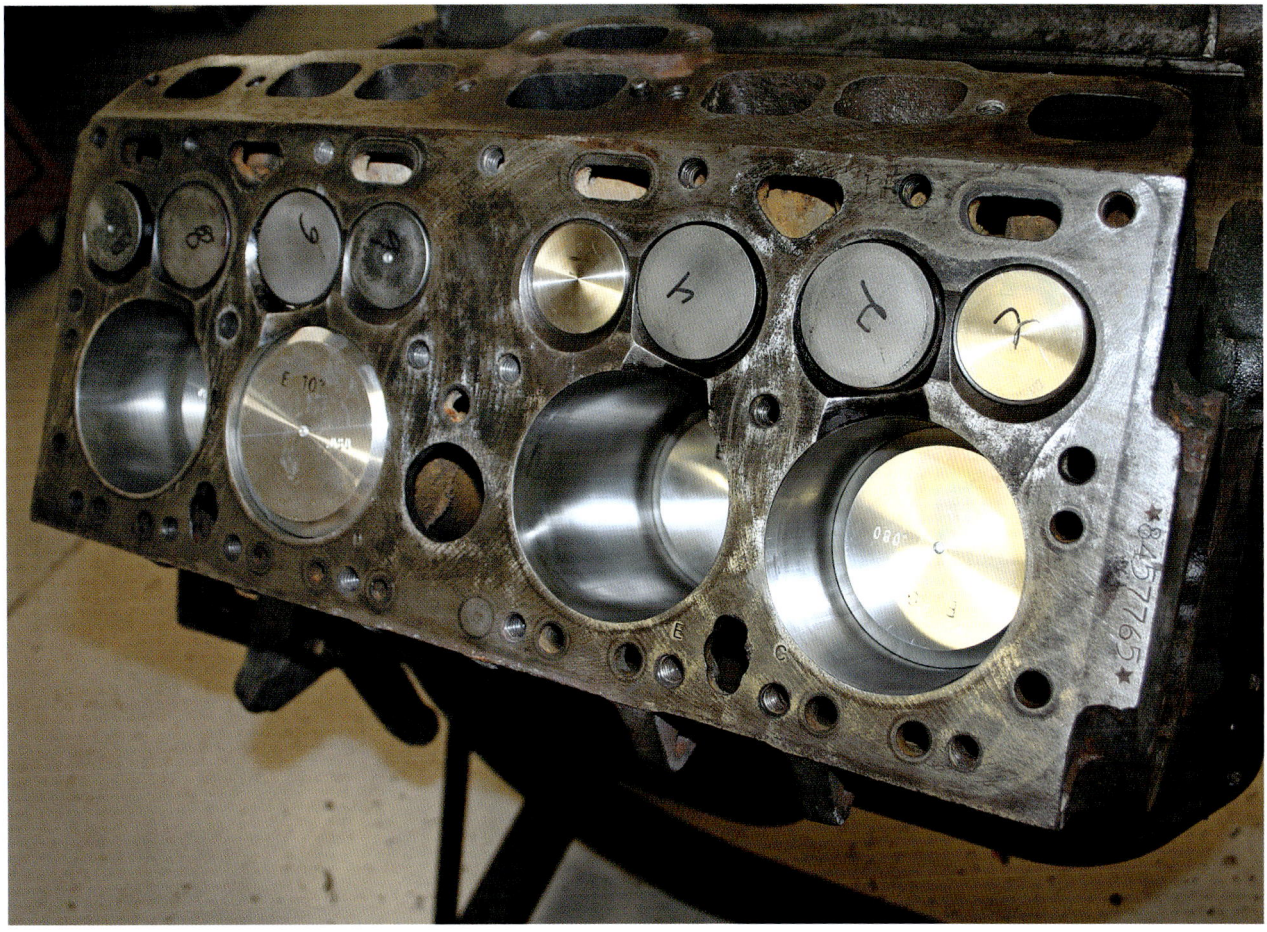

Early flathead engines are easy to spot, especially with the cylinder heads removed. The valves are located in the block, which is why the heads are flat when compared to later overhead-valve designs. This is an early Cadillac flathead V-8, but the popular Fords look similar.

Like the flathead, the early Cadillac and Olds overhead valve engines are experiencing a nostalgic comeback today. They represent a specific period in time, and hot rodders wishing to relive those times are enthusiastically seeking out the engines and speed equipment produced in the post flathead era. Enthusiasts are rediscovering the Ford Y-block, the Buick Nailhead, the early Chrysler/Plymouth/DeSoto Hemi, and other engines once known for their power, and they're finding early aftermarket power upgrades to dress them up with. While crude by today's standards, this vintage speed equipment has nostalgic value, and those wishing to celebrate this historic period value it greatly.

What the hot rodders of the 1950s really wanted was a smaller, lighter V-8 with the breathing benefits and simplicity of the new overhead valve configuration. They would get their wish in 1955 with the release of the legendary small-block Chevrolet V-8.

It was called "The Hot One" right from the start, and even the enthusiastic marketers of the time could have had no idea how good the small-block Chevrolet V-8 would prove to be. The original displacement of 265 cubic inches in 1955 would grow to 283 cubes in 1957, 327 in 1962, 350 in 1968, and a full 400 cubic inches in 1970. Factory power ratings would also increase in an almost linear fashion, and the ongoing development of cylinder heads, camshafts, ignition systems, and both intake and exhaust tracts would give performance enthusiasts plenty to work with in coming years.

The basic design was an excellent one, with no real flaws. It was strong, reliable, easily modified for greater power, and would run cool. Its innovative oiling system would withstand high-rpm levels, the stud-mounted rocker arms offered simple adjustment and stable operation, and the cooling system required no modifications to perform in typical competition situations. This was just as true on the circle track as it was on the drag strip or salt flats.

The fact that Chevrolet produced this engine with no basic design changes from 1955 through 1987 means millions of them are out there, and the price of a rebuildable core engine remains low. Any power-hungry hot rodder can get into a small-block Chevy for little investment, and this key point will keep the so-called "mouse" engine a popular choice long into the future.

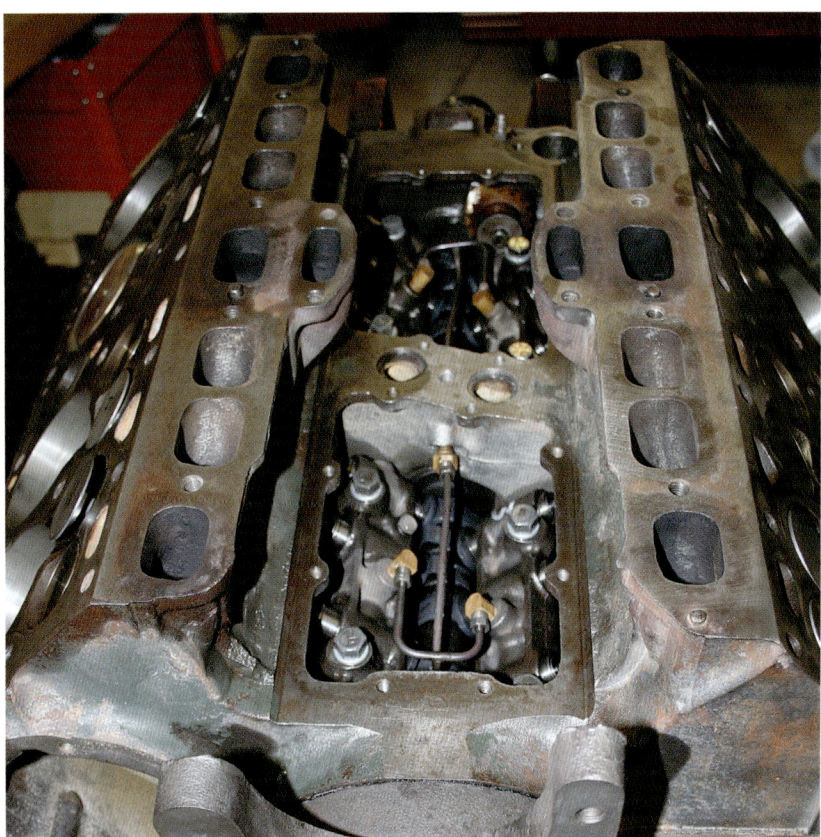

Because the valves are in the block, so are both the intake and exhaust ports. This arrangement was less than optimal for performance, and was a challenge to cast, but the production engines proved to be reliable.

This Ford flathead looks right at home in a traditionally styled hot rod, especially with twin Stromberg carburetors installed. While flatheads like this one didn't make huge power, they made enough to propel lightweight early hot rods. The advent of overhead valve V-8s made flatheads obsolete as performance engines, but they're still fun. It's tough to beat the visual impact of a flathead, and they're surprisingly affordable too.

THE BIG-BLOCKS

As the 1950s became the 1960s, factory engines delivered more power through larger and larger displacements. The public's desire for larger and larger cars, and for those cars to have more and more power, required bigger and stronger powerplants. Horsepower had been a strong selling point for years, and the constant development led to truly amazing results.

Chevrolet's "other" V-8 engine was the 348-cubic-inch "W-block." First offered in passenger cars in 1958, the 348 proved capable of urging on the heavier cars Chevy offered that year. The 348 would gain displacement to 409 cubes in 1961, and it became legendary in 1962 when it proudly made a full 409 flywheel horsepower as delivered from the factory. Further factory development boosted output to 425 horses for both 1963 and 1964.

While the full-size cars carrying these strong factory engines were fast (as evidenced by their record-setting performances in drag racing's Stock and Super Stock categories), when these powerful powerplants were swapped into lighter cars, their legend was forever cast into American automotive lore.

The small-block Chevy (SBC) really redefined the performance level for American V-8s as it evolved throughout the 1960s. This example is decked out in 1960s-era race goodies, like the mechanical fuel injection setup on a Hilborn intake and the Spaulding Flamethrower front-mounted magneto ignition, for use in vintage racing.

Interest in early V-8s, like this 409 Chevy, has prompted development of new components, even though these engines haven't been produced since 1965. These engines, and others like them, can be built to provide solid power and torque in a reliable package. Similar modern development of other early engines, like Ford's flathead and FE and Mopar's early Hemi, offer hot rodders a wide range of choices when deciding how to power their project cars.

This early Chevy W-series big-block engine is fed by a factory tri-power triple two-barrel setup that was offered on the 348-cubic-inch versions of the engine. The valve cover decal claims it's a 409, but it's hard to tell the difference between a 348 and a 409 without pulling a cylinder head and measuring the bore and stroke. Using a factory multicarb setup was a popular way to gain power back when these engines were more common. They're fun to see at car shows today, and they can be made to run well if they're tuned properly.

In 1965, Chevrolet introduced its new Mark IV big-block V-8, at 396 cubic inches. This engine would grow to 427 and then 454 cubes, and would set the bar for all other performance engines to follow. The popularity of these engines prompted aftermarket development for racing and street use, and modified "rat" engines are still popular on the street and at the racetrack. The advanced design of the big-block Chevy, including canted valves, evenly spaced exhaust valves for improved cooling, and the potential to grow to much larger displacements helped it supersede the W engine. The performance potential in the big-block Chevy design made it the standard by which all other domestic V-8s are compared, and the aftermarket options are broad and still being developed for street, marine, and competition use to this day.

The engineering power brokers at the other domestic factories were hardly letting Chevrolet steal the horsepower limelight. Ford's overhead valve V-8s had evolved into the Y-block V-8 by 1954, and the new FE engine family would arrive in the early 1960s. While the Y-block never gained great notoriety as a performance powerplant, the FE would grow into a competent and strong engine still held

The big-block Chevy (BBC) debuted in 1965 at displacements of 396 and 427 cubic inches, and would grow to 454 cubes in 1970. The BBC ushered in a new era of high-performance big-displacement American engines from all of GMs various brands. The Chevy factory offered well-engineered performance components too, like the three two-barrel Holley carb setup on this 1968 427-powered Corvette. Aluminum cylinder heads were also a factory option on big-block Corvettes from 1967 to 1969.

Chrysler's Hemi engines were the kings of the hill at the drag strip; the design was capable of supporting more power than any other of the muscle car era. Recently, Chrysler began producing new versions of the classic 1960s-era Hemi engines in displacements up to 528 cubic inches.

in high regard today. It would power the best of Shelby's Mustangs and Cobra sports cars at 427- and 428-cubic-inch displacements. It would carry Ford's colors into the winner's circle at NASCAR and NHRA events, and power Ford's GT-40 to victory at LeMans. At the peak of its development, it would wear hemispherical heads, displace 429 cubic inches, and become known as the Boss 429. Ford would also release a single overhead camshaft (SOHC) package for the FE block, which created a special following. While the potential of the SOHC 427 package was immense, it proved less than reliable in competition. Ford also developed the basic small-block architecture into a heavier-duty "midblock" version in 1970. Named the Cleveland, it would only be offered until 1974, but the performance potential would prove to be impressive, especially in the Boss version.

Chrysler learned the lessons of overhead valve V-8 power even before Chevrolet had released its first small-block. In the early 1950s, Chrysler's research produced the hemispherical cylinder head, better known as the hemi head. The early 1950s-era Hemi V-8s were distinct and different within each of Chrysler's various internal divisions (DeSoto, Dodge, and Chrysler) and many engine components would not even swap between these similar powerplants. It wasn't until the development of the larger-displacement 354- and 392-cubic-inch-displacement Hemi engines in the late 1950s that things

really started rolling for the Mopar camp. These engines had enough displacement to take advantage of the hemi head's great breathing potential, and had enough strength engineered into their bottom ends to support big power. When superchargers and nitromethane fuel were introduced to racing, the 392 was one of the few factory engines that could handle it without significant modifications.

This durability would carry over into the 426-inch version of the Chrysler Hemi that was introduced in 1964. The Street Hemi debuted with an optional 425 horses when equipped with dual four-barrel carburetors, and much more was hiding inside. With a few well-chosen upgrades, Street Hemi engines could touch 500 horsepower. Mopar offered these same engines across its Dodge and Plymouth nameplates, and this helped make them more attainable to enthusiasts. Better yet, many of the components used in the hemi engine's bottom end were borrowed from the reliable Wedge V-8 program. These 383- and 413-cubic-inch B engines, along with the 426- and 440-cubic-inch raised B engines, lacked the relatively exotic hemi head, but their wedge-shaped combustion chambers were also good, and these big blocks were solid performers.

This point was proven in Mopar's Max Wedge Dodges and Super Stock and Super Commando Plymouth V-8s. In their highest state of development (in 1963–1964), these

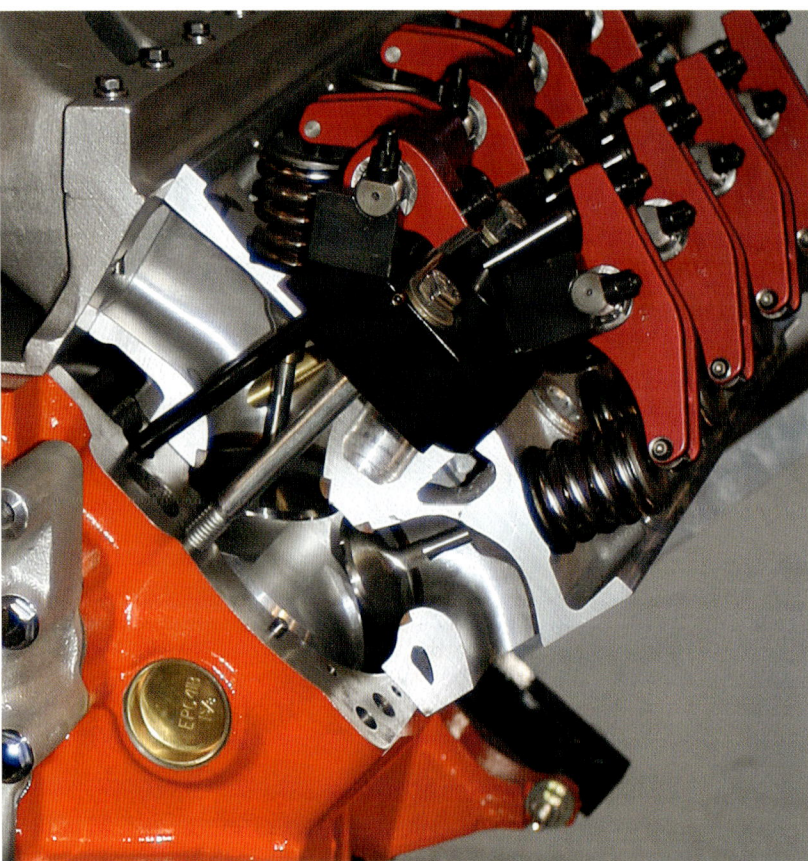

The keys to the Hemi's capabilities are in the cylinder head. Here, we can see how the valves are on opposite sides of the combustion chamber, and that they open on an angle away from the cylinder wall. This lack of shrouding allows for big valve sizes and generous port diameters capable of moving lots of fuel and air in and out. By increasing the cubic inches, the large ports became more efficient too. Adding a supercharger was a natural progression, and it wasn't long after its introduction that the Hemi was choice of Top Fuel teams everywhere. The extensive and complex valve train is expensive, as are many of the Hemi's unique parts.

426-cube engines came with a factory aluminum dual-quad cross-ram intake manifold, a cast-iron high-flow exhaust manifold, 12.5:1 compression ratio, solid lifters, a high-rise camshaft, and an advertised 425 horses. Some of these cars still hold world records in NHRA Super Stock classes. While Pontiac may lay claim to launching the muscle car era with its '64 GTO, there's little question some muscular Mopars of the time were already capable of giving them a good run for the title.

Later, hemi and wedge engines would gain more widespread availability and recognition in later years, as Chrysler Corporation fully embraced the muscle car. Offering their biggest and best engines as options in their B-body, A-body, and, eventually, E-body chassis would make for some truly legendary machines. Keeping more sedate versions of the B and raised B engine available throughout the 1970s, and continually refining their fine small-block A engines well into the 1980s, made the Chrysler small- and big-block powerplants popular with racers and enthusiasts.

The muscle car era ran strongest from the mid-1960s to the early 1970s, and every nonluxury domestic auto manufacturer was offering relatively small cars with relatively large engines in some form or another. While Chevrolet's efforts have been mentioned, the engineers at Buick, Oldsmobile, and Pontiac were all offering some amazing cars as well.

Starting in 1964, special sport packages were offered on the midsize GM line, and these included increased-horsepower options. Buick's Skylark had the Gran Sport (GS), Oldsmobile had the 442 (designating a four-barrel, four-speed, dual exhaust package) for its Cutlass, and Pontiac had the aforementioned GTO, which was based on the Tempest/LeMans and delivered with a 389-cube V-8. Pontiac aggressively marketed its GTO, and this resulted in its great success. Only Ford's Mustang would make a bigger splash with enthusiastic new car buyers than the GTO, and there was little question the GTO was faster. With 100 more cubic inches to work with (389 versus Ford's 289), the Pontiac could lay waste to an early Mustang at any stoplight.

Interestingly, Buick, Oldsmobile, and Pontiac would continue development on their respective big V-8s until each offered a 455-cubic-inch version. While none of these 455s had any relation to each other, and each was a distinct and different engine design, they were similar in performance. Big torque numbers, relatively low rpm redlines, and responsiveness to aftermarket bolt-ons made any 455 a terrific street performance engine. Each of the GM 455 V-8s has its own quirks; all have proven themselves as capable and durable powerplants when built right.

While Hemis ruled the track, the wedge-headed engines got their fair share of attention from the factory too. Just before the Street Hemi was introduced, the 426-ci Wedge engines were offered with dual four-barrel carbs on a cross-ram intake manifold. These engines could be had with a 12.5:1 compression ratio, and many preferred them to the more complex and expensive Hemis.

THE SMALL-BLOCKS

While the big-inch engines began their development early, as time passed it became evident that small-block V-8s would also become big players in the hot rodding marketplace. The small-block Chevy set the bar, but well-refined small-block engines from Ford and Mopar would make a big splash in their own right. Ford would place its bets on the Windsor-based V-8s, which began with the 260- and 289-inch engines in the early 1960s. These engines began to shine as Boss 302s in the early 1970s, and were the basis of the 5.0-liter engine program that took off in the 1980s, thriving even as electronic fuel injection and computer controls were added. It wasn't until Ford launched its Modular V-8 program in 1997 that the Windsor was finally relieved of duty. This means that millions of potentially potent Ford V-8s are out there, offering affordable options to enthusiasts. Like the small-block Chevy, parts for the small-block Ford Windsor are readily available everywhere, and will be for a long time to come.

Every mainstream (nonluxury) American manufacturer eventually had a strong small-block package in a sporty body. Buick and Olds had well-developed small-block 350-inch engines, Pontiac's mid-block architecture allowed them to offer displacements from 326 to 455 cubes in the same basic crankcase. Even American Motors had refined their outstanding small-block design to be a real contender in its 343-, 390-, and 401-cubic-inch displacements.

In a similar vein, Mopar developed a great small-block and offered it in a wide range of cars and trucks for decades. Typically seen in 273-, 318-, 340-, and 360-inch displacements, the small-block Mopar found success in Chrysler's popular A-body platform. These smaller chassis accepted the small-block with ease, and their light weight made the small-block Dusters and Demons of the early 1970s real contenders. Dodge trucks gained a reputation as capable off-road machines when equipped with the later 360-inch engines.

The success of the great American V-8s had a major impact on the development of their modern replacements. GM's LS-series, Ford's Modular V-8s, and Chrysler's new Hemi all owe some debt to their predecessors. This may be most evident in GMs case, as its current family of V-8s maintains external dimensions (and a commitment to pushrods) that are strikingly similar to the 1955 version. Engines in the LS-series of V-8s excel in every comparison to the earlier version (from strength to efficiency to lighter weight), and these new generation engines command a relatively hefty price. Their excellence justifies the cost, but for those wanting to make

Ford's Windsor family of small-block V-8s was born in the early 1960s, and was offered in various forms into the 1990s. This long life means a wide range of great parts were developed for them, and you can still get a reliable Windsor-based engine new from Ford Racing or any of a wide range of premium builders. This beautiful 450-horse, 351-cubic-inch example is from Roush Racing.

The introduction
of the GM LS-
series V-8 in 1997
represented the latest
generation of the small-block Chevy
V-8. Its power potential, refinement, and
durability have already made it a favorite among hot
rodders. Shown is a factory crate version of the LS outfitted with a
carbureted intake manifold for use in ASA stock car racing.
Photo courtesy of General Motors

Ford engineered an overhead cam V-8 to power their modern performance cars, and they offer outstanding factory-engineered powerplants through their crate engine program as well. The supercharged versions of these crate engines produce up to 600 horsepower. Shown is a more sedate 4.6-liter version.

Drawing on a historical name and chamber design, Chrysler's new version of the Hemi V-8 brings efficient power in a small-block sized package. While it lacks the huge dimensions of its 1960s-era namesake, Mopar enthusiasts are finding great power potential in the new Hemi, especially in the larger 6.1-liter versions.
Photo courtesy of Chrysler Corporation

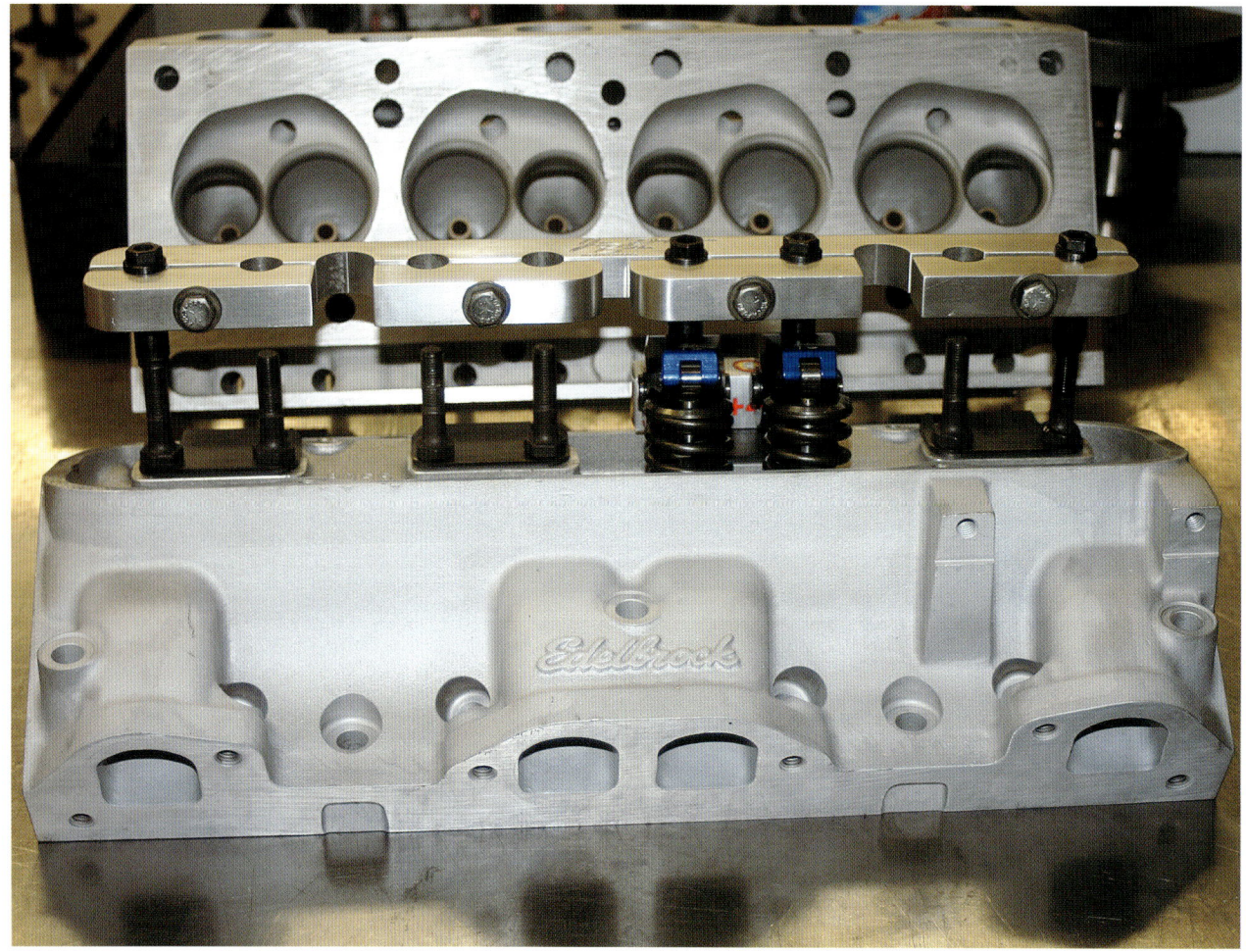

The traditional Pontiac is a good example of a vintage American V-8 that is receiving renewed attention, with several aftermarket manufacturers bringing new cylinder head designs to market. Shown is a pair of Edelbrock heads for the Pontiac. This particular set has been worked over for increased power production by the Pontiac specialists at Butler Performance Products in Leoma, Tennessee. Due to parts like these, these classic engines can produce much more power than they did in their heyday.

respectable power on a budget, it's still tough to beat the old small-block Chevy.

The folks at Ford chose to go with an overhead cam design for the Modular V-8, and its subsequent success and power capabilities prove this was a good move. While its cam covers are wide enough to scare many engine swappers, the 4.6- and 5.3-liter engines make plenty of power. Ford's choice to offer two-, three-, and four-valve versions of the V-8 cylinder head opened more doors, giving hot rodders the ability to match cylinder head and valve configuration to a particular application. When the largest-displacement bottom end is paired with the four-valve cylinder head and topped with a supercharger, Ford engineers have been able to offer 500 reliable horses from these relatively small-displacement engines.

Chrysler's new Hemi engine borrows more than just its name from its predecessor. The new Hemi is technically a small-block, with its largest factory displacement offering limited to 6.1 liters at the time of this writing. While the

hemispherical chamber is significantly shallower on the new engine, it's still there to some degree. Chrysler engineers have made the new Hemi a wonderful option for cars and trucks alike, with respectable fuel economy figures to balance solid horsepower numbers. As the new Dodge Challenger heads to market, the romance of a true Hemi-powered Mopar muscle car is once again a reality. What future development may occur to the new Hemi remains to be seen, but its selection as the powerplant for the high-profile Challenger SRT is encouraging. These new American V-8s, with their computer controls and high-efficiency designs, will continue to be great choices for future hot rods.

Like the new engines being offered from the great American auto manufacturers, vintage engines are also improving. The automotive aftermarket is full of enthusiasts who are always looking for ways to gain more usable power or durability, so new parts continue to be developed. While some of the great American V-8s have been out of production for decades, many new parts for them have come to market in

Renewed interest in vintage American V-8s has prompted development of new parts for many engines not produced for decades. While production of the Chevy 409 ended in 1965, the aluminum cylinder heads, intake manifold, water pump, and valve covers on this 409 are newly developed Edelbrock parts.

just the last few years. The reasons for this are varied, but the combination of continued enthusiast demand and computer-aided manufacturing have helped, especially in the area of cylinder heads.

For many years, engine builders—certainly those working with the less-mainstream engines—were limited to the factory-produced cylinder head castings. Race-specific heads had been built (by both factories and the aftermarket) for the small- and big-block Chevy, Ford, and the small-block Chrysler and Hemi. More recently, well-designed aftermarket cylinder heads for the traditional Pontiac, Oldsmobile, and Buick V-8s have come to market.

Additionally, aftermarket high-performance aluminum cylinder heads are now offered for AMC V-8 and even Chevy's 348/409 engines. These aftermarket heads are not mere copies of factory heads—they've been redesigned for greater efficiency from the ground up. Improvements to port design, chamber shape, and valves are usually incorporated, and fans of these vintage engines are enjoying a newfound renaissance. As good as these engines may have been when they were new, they can be made even better now.

Research into camshaft science and manufacturing has never stopped, and the latest generations of camshaft profiles have proven to be effective in new and vintage American V-8s. When teamed with new cylinder heads, it's not hard to see how once-forgotten engines are finding new life and new power peaks.

There's little question that advances in automotive electronics have helped the new generations of V-8 reach new levels of efficiency. Electronic fuel injection (EFI) and computer-controlled ignition systems are capable of tuning these engines on the fly, and when teamed with cutting-edge valve control systems, impressive fuel economy levels are also possible. Some enterprising enthusiasts have adapted EFI to older engines that were never offered with it, and the benefits

were recognized immediately. In a similar vein, modern computer-controlled ignition systems are becoming the norm on today's engines. These distributorless designs fire the spark plugs when the computer tells them to, and their ignition timing is infinitely variable. Based on sensor feedback, the timing is advanced or retarded to provide the most efficient performance in a given situation. Unlike traditional mechanical means of advancing and retarding timing—weights and springs controlled by centrifugal force or vacuum advance mechanisms—late-model direct ignition systems (DIS) have programmed timing curves that are incredibly accurate. They can account for supercharger boost levels, nitrous oxide injection, high altitude use, and other factors traditional ignition distributors cannot accommodate.

While it's not yet possible to adapt modern variable valve systems or cylinder shutoff technology to vintage engines, it would not be shocking for something to be developed in the future to accomplish this. As technology advances and becomes more accessible and affordable, it will continue to be adapted to older engine designs, and they will continue to evolve along with the newer engines, whose design and development they influenced. It seems only fair, doesn't it?

Late-model rides can gain power and performance without even getting dirty under the hood. Power programmers (like this one from Hypertech) change the settings on factory computers to provide more power and rpm capability. If you swap an entire EFI setup from a stock late-model computer-controlled vehicle (or your ride happens to be just such a vehicle), it may be possible to use a programmer like this one to wake it up a bit.

Chapter 2
The Engine Block

Every powerful V-8 engine must have a solid foundation, and this begins with the engine block. If you're fortunate enough to be starting with an aftermarket block, your project engine will benefit from the improved strength such blocks provide. Most engine projects are based on rebuilding a factory block, and luckily for enthusiasts, the overwhelming majority of American factory V-8 blocks are suitable for performance use. With some basic upgrades and a machinist's blessing, virtually any factory block can support double the power it shipped with when new. To understand what modifications to make and how capable your block of choice is, let's first understand what an engine block has to do.

Shown is a factory 305-ci SBC block next to a GM Bow Tie block designed for high-performance use. The obvious differences from this angle include the much heavier main bearing caps on the Bow Tie block. The increased strength in this critical area allows the Bow Tie block to support much higher power levels and increased durability in high-rpm or high-endurance applications.

When we look into the lifter valleys of the stock 305 (left) and the Bow Tie block (right), more differences become apparent. A combination of very aggressive cam lobes and high valve spring pressures common in high-performance engine designs puts a lot of stress on the lifter bore area. The additional material in the Bow Tie block will withstand these stresses; the 305 may not.

Another significant difference between production line blocks and aftermarket high-performance blocks exists in the cylinder head deck mating surface. The high cylinder pressures common in high-performance engines means the stress on this surface is high. A thicker deck is less likely to fail, and offers a more stable platform for the cylinder heads. The 305 deck is relatively thin (top) when compared to the Bow Tie version of the same block. Luckily for fans of vintage V-8s, they typically had very thick deck surfaces as manufactured at the factory.

The engine block is the basis of the engine, and is therefore the foundation for any engine build. A thorough inspection of critical areas will ensure the block is capable of withstanding the forces of a high-performance existence. Casting flaws, damage, and excessive wear are all reasons a block could be rejected. A good mechanist will know what to inspect, and will have the proper tools to accomplish such an inspection.

Engine blocks are also called crankcases, and this is a good descriptor. The block does house the crankshaft, along with the pistons, connecting rods, and (in non-overhead cam applications) the camshaft. The stresses associated with containing the energy of these moving parts and the heat generated by combustion explains why engine blocks are typically such heavy castings. They need to be strong, especially in the main bearing area and along the deck surface where the cylinder heads will bolt up. The fasteners used in these areas are especially critical, and a typical performance engine build includes an upgrade to top-quality hardware.

Just as important as good hardware is the condition of the holes the hardware screws into. Checking for thread condition is part of your machinist's job, and inspecting the entire block in preparation for high-performance use should be part of your plan from the beginning. One of the critical inspections that the machine shop should perform is to look for core shift in the cylinder bores. If the bore sleeves were not centered in their core boxes when the block was first manufactured, it may limit the block's capability. The core shift and cylinder wall thickness can all be inspected by a competent machinist before any other work is done, and this can save you time and money before you invest heavily in a bad block.

Here a crankshaft is being Magnafluxed to check for cracks. In this procedure, a steel part is soaked in fluid containing fine metal particles and green dye. Low current is then passed through the part, and any cracks in the part will retain the dye, which can be inspected using a blacklight. This should be a standard procedure at the machine shop you choose.

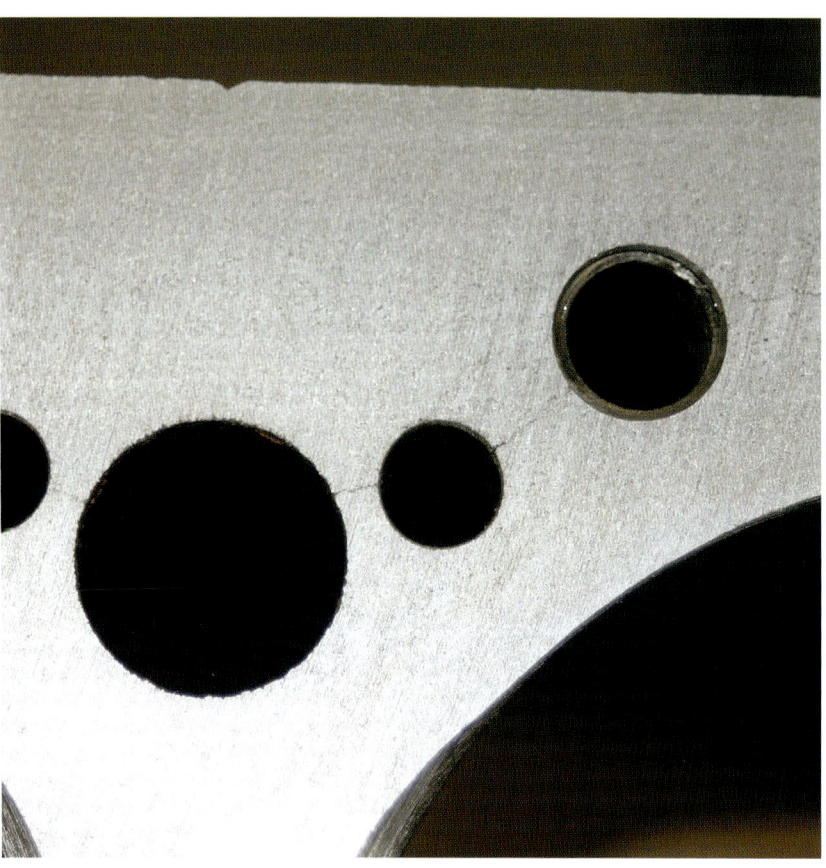

This is why blocks need to be inspected carfully! You can see a crack extending from hole to hole across the deck on this block. This would have caused huge problems if it hadn't been spotted during the inspection process.

Here's another crack in a block; this time in the lifter valley area. The red powder you see is metallic, and when the block is magnetized it is drawn into any cracks. It's an effective inspection method that makes cracks like this easy to spot.

Another issue to be aware of is block core shift. This happens when the castings move a bit while being manufactured, and the machining doesn't end up precisely centered. The offset nature of the lifter bore hole and the camshaft tunnel shown on these blocks may be severe enough to exclude them as potential high-performance canidates. While core shift like this may present no issues in a stock build, the added stress of high-performance operation could lead to breakage in these critical areas.

Core shift can also affect the thickness of the cylinder bores, which becomes very critical when an overbore is part of the plan. To avoid any potential issues, the thickness of the bore can be checked using this sonic gauge. Every bore is checked in four different directions, and if any cylinder wall is determined to be too thin (after the required overbore deemed nessecary), the block could be rejected.

A full machine shop treatment should include resurfacing of all the flat surfaces on the block. The deck surfaces should be measured not only for flatness, but also to ensure they are equal distances from the crankshaft centerline. This means both of the cylinder heads will be the same distance from the bottom of the block, and also that all the engine's cylinder bores will have the exact same height. This becomes more important when developing custom pistons and determining compression height and compression ratio, but it begins with the machining on the bare block.

The machined surfaces inside the block are also of critical importance. The cam tunnel, crankshaft bore, and all the cylinder bores should be checked for size and concentricity (roundness). While these machine shop practices can be considered "standard procedure" for any performance engine rebuild, some additional procedures have been developed in recent years. For example, it has become popular to bolt torque plates in place during the boring of the block. This simulates the stresses placed upon the block when the heads are bolted on. This makes for a better piston ring seal, which is a major point in any rebuild.

In a similar vein, many machinists are experimenting with hot honing. In this process, liquid is circulated through the block while it is being machined. The liquid is typically heated to around 200 degrees Fahrenheit to match a typical operating temperature. If we machine a block with torque plates installed and have it heated to 200 degrees, it truly replicates the conditions the block will see while it's running. If the machine work is accomplished under these conditions, the block's internal dimensions should be closer to design specifications, and the engine should perform better.

The engine block is stressed most in the area of the main bearing caps. This is easily understood when we look at the design of a typical engine block and what it's being asked to do. When the piston is forced down the cylinder bore, pressure is transferred through the connecting rod to the crankshaft journal. As the crankshaft rotates, the pressure load is transferred to the main bearing in the block. While the block casting itself is plenty strong, the main bearing cap bolted to it is typically the weakest point. If any investment is going to made to beef up your block, this is where it should be. While upgraded main bearing cap bolts (or studs) are recommended for any performance engine, upgrading the main bearing caps themselves is also worth looking into, especially prior to having the block fully machined. This is common in many popular engines where the manufacturer provided two main bearing cap bolts, and you'd like the additional strength

A resurfacing machine provides a level surface by removing a predetermined amount of metal from the part in question. A cylinder head is loaded in the resurfacer in this photo, but the same machine is used on engine blocks. The spinning, horizontal head shaves the surface gradually and cleanly. This way, the block and heads will all be uniform in height and surface condition for the best-possible seal and identical performance from each side of the V-8.

You can clearly see the steel cylinder sleeves used in this aluminum GM LS-series engine block in this photo. The use of cylinder sleeves to repair damaged blocks was long looked down upon as a cheap repair, but the increased use of sleeves in factory aluminum blocks is changing this mindset. With better-quality sleeves being developed, owners of both aluminum and iron blocks can choose to use sleeves without fear.

Here, a block is shown in the cylinder bore machine. You can see the honing plate is torqued in place to simulate the stresses of an installed cylinder head. The arm moves up and down as the honing stones spin to provide a cross-hatch finish.

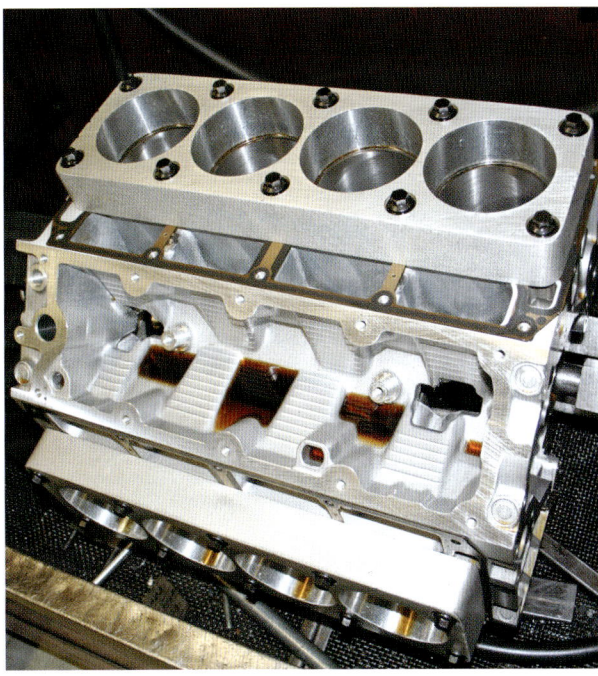

While many builders have been using torque plates for years, it's even more critical with modern aluminum blocks. Aluminum moves under pressure more than steel or cast iron, so the effect of the cylinder heads being bolted on an alloy block will be greater than with an iron block. Having both torque plates installed while machining an aluminum block is critical to ensuring the bores are round when assembled. Bringing the temperature of the machining lubricant up to 200 degrees or so (operating temperature) also helps, as the aluminum will also expand with heat.

four main bearing cap bolts can provide. Opinions vary on whether upgrading to a four-bolt main bearing cap design is actually stronger than a two-bolt arrangement, but there's little doubt that upgrading to a splayed four-bolt design will increase bottom-end strength. Personally, I've never seen a four-bolt main bearing cap fail, and I will always choose a four-bolt block over a two-bolt block. If I have a two-bolt main bearing block I'd like to upgrade, I will always choose a splayed-bolt aftermarket cap.

If you need the additional strength aftermarket bearing caps can provide, you'll have to have them machined to fit. The time to do this is before the rest of the block is machined, because the block should be machined with the main bearing caps installed and the fasteners torqued to specifications. Like the torque plates that simulate the cylinder heads, having the main bearing caps installed and tightened will replicate the torque and stress the block sees when it is fully assembled.

If a four-bolt main bearing cap upgrade is not available for your engine, there are still ways to increase the strength of the bottom end. Strap caps are one way. These simple metal straps fit across the two main bearing cap bolts and offer additional strength. Another method is through a stud girdle (or halo), which typically fits over the main bearing

caps and sandwiches between the oil pan and the bottom of the block. It transfers stress out to the oil pan rail and offers additional support to the main bearing caps.

Another job the engine block must do is to both contain and circulate coolant to remove the heat generated by the engine.

Some domestic V-8 blocks circulate coolant and control heat better than others, but most do a pretty good job. Still, it makes very good sense to not only have your block fully hot tanked and flushed to remove old scale and rust from the coolant passages, but also to physically inspect all of these passages yourself. Sometimes, casting flash from the factory or rust scale may restrict a coolant passage. If casting imperfections are found, they can be ground down to improve coolant flow through the block. Rust scale that has broken off may have to be picked out by hand or blown out with an air gun. Making sure the coolant has a clear and simple path around the cylinder bores, up into the cylinder heads, and both to and from the water pump requires only a simple visual check, and it's well worth the time to accomplish. The passages through the block's deck surface up into the cylinder head are of particular concern, and these should be cleaned by hand and reinspected after the machining is complete.

For additional support, some performance-based engines rely on four-bolt main bearing caps as opposed to the two-bolt caps typically found on most engines. While two-bolt caps have proven adequate in many high-performance street applications, four-bolt caps offer a higher level of security and should be considered for any engine destined for a life of extreme duty. This is an aftermarket four-bolt main bearing cap offered by Milodon.

Shown from top to bottom are a factory two-bolt main bearing cap, a factory four-bolt main bearing cap, and an aftermarket billet steel splayed four-bolt main bearing cap. It's easy to see how the greater support offered by the four-bolt caps is preferred for true performance applications. The outer splayed bolts used in the billet piece spread the load out over a greater area, lessening the stress directly below the bearing cap. This helps support even greater power levels.

Some engines, like race-level Chrysler Hemis and even the new GM LS series (shown), have six-bolt main bearing caps. The additional bolts pass through the sides of the block into the main bearing cap for additional security. Such arrangements are called cross-bolt main bearings.

A used block being rebuilt will need to be thoroughly cleaned, and special machines exist to accomplish the task. A combination of oven baking and machine washing (shown) will effectively remove all of the oil, grease, rust, and scale buildup the block accumulated over its life. This makes it easy to inspect, handle, and work on. After all the machining processes are complete, the block will be washed again prior to assembly to ensure cleanliness.

While it's normal for some corrosion to build up in the water jackets, too much rust buildup will limit coolant flow and cause cooling problems. Today's block cleaning procedures do a great job of removing this corrosion, and restoring the cooling capability. You can check the level of corrosion by peeking in through a removed freeze plug hole. This block shows typical rust levels.

Block filler (like this product from Hard Blok) is a cement-like substance that is poured into the water jackets of a block to add structural integrity to the critical bottom end of the engine. Naturally, this means less coolant capacity within the engine block, which must be taken into consideration. Depending upon the block design, it may be possible to add a limited amount of block filler to gain required stability without sacrificing too much cooling capability.

In some applications, block filler is recommended to add stability and strength to the bottom of the block casting. This is typically done by pouring a concrete-like substance into the block's cooling passages, usually through the freeze plug holes in the side of the block. Many blocks can still circulate enough coolant to keep the engine from overheating during normal use after using block filler, but not all of them. Physically larger blocks normally hold more water, and are better suited to partial filling, but it's best to rely on the advice of a brand-specific expert in this area. Any block that is partially filled should have the rest of the cooling system upgraded to ensure the best-possible cooling potential. For some engines, like big-block Buicks, partially filling the block is an accepted and proven way to add strength and stability to the bottom end while maintaining street cooling capabilities. For the majority of domestic V-8 blocks, filling should only be a consideration in the most extreme applications, and is typically not a requirement for street-driven engines. A block with lessened coolant capacity will be more sensitive to overheating.

Similarly, the engine's oiling system relies on passageways through the block, and these must also be inspected and cleaned thoroughly. In most cases, the oil pump mounts directly to the block, and the passageways between the pump and the block must be cleaned, smoothed, and checked for proper alignment before assembly. The importance of

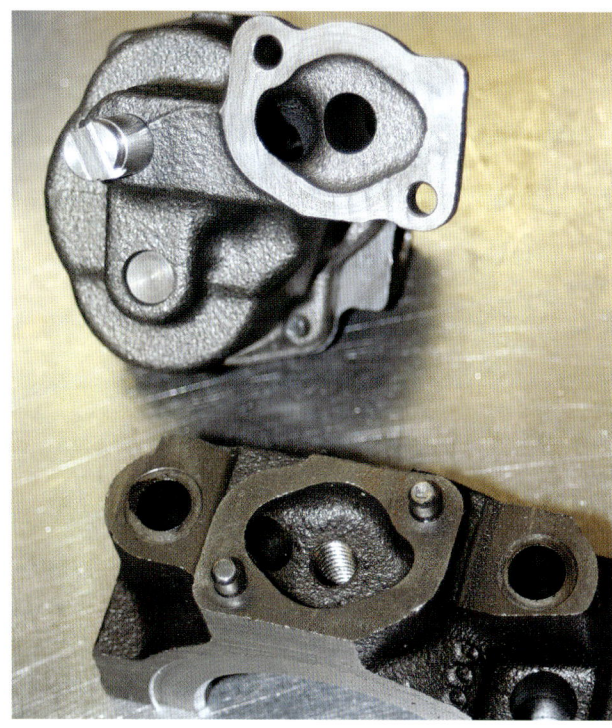

Both the oil pump housing and its mating surface (in this case, on the rear main bearing cap) are typically rough castings. Since precious oil enters the engine through this passage, it makes sense to clean up the rough surface to aid lubricant flow.

Some cleanup work with a cartridge roll is all it takes to make a difference. Simply smoothing out the rough surface of the casting guarantees no casting flash will restrict the flow, and the oil pump can deliver to its full capacity.

An engine block's oil passages must be clean and clear to provide precious lubricant in critical areas. Shown is a main bearing oil passage in a GM LS1 engine block. This hole will line up with a similar hole in the main bearing, and both holes should be sized to match and deburred to guarantee smooth and steady oil flow across the main bearing surface, where it meets the crankshaft's main bearing journal.

the engine's lubrication system is obvious and cannot be overstated, and block preparation is a big part of this.

All of the block's oil passages and galleys need to be thoroughly cleaned and visually inspected for obstructions. It's common to use long rifle-cleaning brushes in the passageways alongside the camshaft tunnels that are common in domestic V-8s. Plugs on either end of the block offer access to these galleys. The job of these galleys is to feed oil to the cam and lifters, so they need to be spotless and clear. The plugs themselves are also critical, as the oil passages typically intersect at the end of the block, close to where the plugs are installed. If a longer-than-stock oil galley plug is used, it may partially or completely block off a critical oil passageway, which can only lead to trouble. Carefully note the exact dimensions of the factory plugs as they are removed, and make sure the replacement parts are identical to the originals. Many engine builders confidently reuse the factory plugs to prevent problems in this area, and if the factory parts are of good quality and in good condition, this is fine. I typically use some sealant or liquid thread locker to hold plugs in position and prevent leakage. By being careful and making a point to double-check every block plug, it's possible to avoid problems down the road.

Because so much of the oiling system flows through the block, it's important to discuss oiling system design and

upgrades prior to discussing assembly. If any upgrades or modifications to the oiling system are planned or required, it makes sense to get these done before the assembly process begins. Any grinding or drilling should be done before the block is cleaned for assembly, so it pays to research lubrication modifications early.

Typical American V-8 oiling systems draw oil from the pan, through the oil pump, then through the filter before pressurizing the block. The oil is fed to the crankshaft and/or camshaft, and then up to the valvetrain. Oiling the rockers and valve springs is typically accomplished through either hollow pushrods or hollow rocker shaft assemblies.

Inspecting and cleaning all of the passageways that the oil travels through is a basic tenet of engine building, but the block is typically not given the attention it deserves. Grinding and smoothing the passageways where the oil enters the block is a good idea, as is "port matching" (making sure the opposing passages of each component match in size and shape) the oil pump to this passageway. The engine's oil passages should all be smooth, and if any rough casting surfaces exist in oil passageways, efforts should be made to smooth these out.

Once you've ensured the pressurized side of the oil system is ready to go, it's time to consider the oil recovery (drainback) system. Because this function uses no moving parts, few consider oil drainback when building engines. But

As effective as modern block cleaning machinery is, the block's internal oil passages still require a thorough scrubbing to ensure they are completely clean. To do this job effectively, long rifle brushes are used. They are capable of cleaning the entire length of the oil galley from front to rear, and can also be used on the passages feeding the main bearing caps.

The oil galley plugs are more important than they seem. Many factories used simple press-in plugs for these holes, and performance applications require the improved security of screw-in plugs with sealant applied to hold them in place. The holes are tapped, and then the screw-in plugs can be installed. The issues arise if the plugs are screwed too deeply into the block, since they can potentially block the flow of oil in some cases, like in the block shown. In these situations, the depth to the oil transfer hole is carefully measured, and the appropriate-sized plug is installed.

it's just as important to consider as the pressure side, and is even more important in a performance application where high-rpm use is planned. Higher-rpm levels mean more oil is pumped to the top end of the engine, so more oil will have to drain back into the oil pan. Many builders choose to use high-volume (HV) or high-pressure (HP) oil pumps in their performance engines, and these move oil to the top of the engine even more aggressively than the factory pumps. Getting the oil to drain back down efficiently becomes a major consideration, and knowing where to look and what to do is pretty basic. This actually begins in the cylinder heads, but block work is a big part of it, too.

Oil will drain from the heads into the valley between them. Most V-8s are designed to direct the oil flow down the front of the valley, the rear of the valley, and around the camshaft. The oil flowing down the front of the valley typically flows behind the timing chain and into the front of the oil pan. The oil going down the rear of the valley will pass around the rear of the spinning crank and into the rear of the oil pan. The oil draining back around the camshaft will fall down around the spinning crankshaft and into the oil pan. Smoothing these pathways is all that is typically required, but some modifications can help here.

The oil that drains around the camshaft and into the spinning crankshaft acts like water being thrown on a fan. It adds to the weight of the reciprocating assembly and slows it down. Naturally, these effects are minor, but they can be avoided by adding a simple sheet metal tunnel under the camshaft to catch this draining oil, direct it to either the front or rear of the block, and keep it out of the spinning crankshaft.

Once the block has been fully cleaned, prepped, and machined, it should be deburred. The machining process will leave sharp edges along all of the freshly cut surfaces. Take the time to file down these sharp edges with a hand file or hand grinder. Simply drawing it along the sharp 90-degree edge at a 45-degree angle to smooth it out is all that's required. This removes the hard edges where a crack could begin, and it makes the block much easier to handle during the assembly and installation processes. I also carefully clean the edges along the top and bottom of the cylinders; care must be taken not to touch the refinished bore surface. Spending some time on this prior to assembly really saves headaches later.

After a final cleaning with soap and water, dry the block with clean, lint-free paper towels and apply a bit of oil to the bores to keep rust from forming. Now, the freeze plugs can be installed. These plugs may be made of steel, aluminum, or brass; I have no preference on the material. They should be installed carefully and must be level. I always run a small bead of sealant around the outside edge of the freeze plug to prevent leaks. In extreme situations, I've installed small screws on the edge of the plug (with the fastener head overlapping the edge of the plug) to secure it in place. Naturally, this means it won't function as a safety against overheating or freezing any longer, but in racing situations this is the norm. These fasteners must also be sealed, and great care must be taken when drilling and tapping the holes for these security screws. I've even safety-wired them for absolute fastener security in full-race applications.

Finally, after all of the work on the block has been accomplished, the block can be painted. I normally wipe

Most of the engine oil returns to the oil pan by draining down through the lifter valley. The passages designed to accommodate oil drainback are cast, and can be made more efficient with a little grinding work. The passage on the left has been enlarged and its edges have been smoothed out. The passage on the right hasn't been touched yet. Note the sharp, raised edge around the untouched passage. This will inhibit drainback.

The sharp edges on factory blocks should be deburred to limit any sharp edges. Not only can cracks develop from sharp edges, but handling the block is also a lot easier. If a deburred block is machined for a rebuild, sharp edges will once again require attention along all the edges of newly machined surfaces, like the cylinder head deck on this block. You can see the effect of the grinder along the edge near the bottom of the photo.

Oil restrictors like these, installed at the back of the block in the cam galleys, limit how much oil flow is available to the pushrods (and therefore the rocker arms). By limiting oil flow to the top of the engine, more oil is available to the crankshaft and connecting rods. These are not recommended in most cases, as a well-designed lubrication and drainback system should make them unnecessary. They used to be common in Oldsmobile V-8s used for extended high-rpm use (like in ski boats, where big-inch Olds engines were once very popular). Modern engine builders focus on effective drainback routes and good windage control, in addition to grooved and/or coated bearings.

the entire exterior block surface down with lacquer thinner and let it air dry before applying tape to all the machined surfaces that will not be painted. I normally apply two coats of primer and two coats of color, and I prefer the high-temp engine paint developed for this purpose. A clean, freshly painted engine not only looks great, but makes it easy to spot any leaks when the engine is newly installed. Any engine worth building is worth painting, and this is another step than is worth the time and effort to do well.

After removing the tape, the block's machined surfaces should be wiped down once again prior to assembly. If the engine won't be assembled right away, wipe the machined surfaces down with a thin coat of oil and wrap the block in plastic to prevent rust from starting.

HV and HP pumps are available for many popular domestic V-8s. High-volume pumps typically have the capability to move larger quantities of oil by having taller pump gears and larger passages to move the oil through. When an engine is run at high rpm for extended periods, it may require a larger volume of oil than the factory engineered for the stock application. The accepted norm for oil pressure has always been 10 psi of oil pressure for every 1,000 rpm, but research into this has found little reason for more than 40–50 psi of oil pressure at any rpm in most V-8s. Knowing

this, and also understanding that it takes a fair amount of power to turn the oil pump, limiting oil pressure to 40–50 psi is adequate and prudent. For this reason, it's much easier for me to recommend a HV pump over a HP unit in anything but a race-only application. Running higher-than-required oil pressure serves no real purpose, and may push too much oil up to the valvetrain. This can outpace the engine's capability to drain that oil back, resulting in the valve covers filling up with oil, and potentially starving the pump pickup because of a lack of oil in the pan. In the past, this has been addressed by installing oil restrictors in the passages feeding oil to the engine's valvetrain to keep oil available to the main bearings and camshaft.

Proper research into the lubrication requirements of your project engine will reveal any special areas of concern. For example, big-block Buick oil pumps require very specific internal clearances to work reliably at higher rpm levels. You can learn of any lubrication system concerns in your engine of choice by talking to brand specialists and racers running similar equipment. Addressing any make-specific concerns, along with proper block preparation and attention to oil drainback pathways, should help you avoid having to restrict oil flow. I'm not a fan of oil restrictors, and I feel that they should not be required in a well-designed

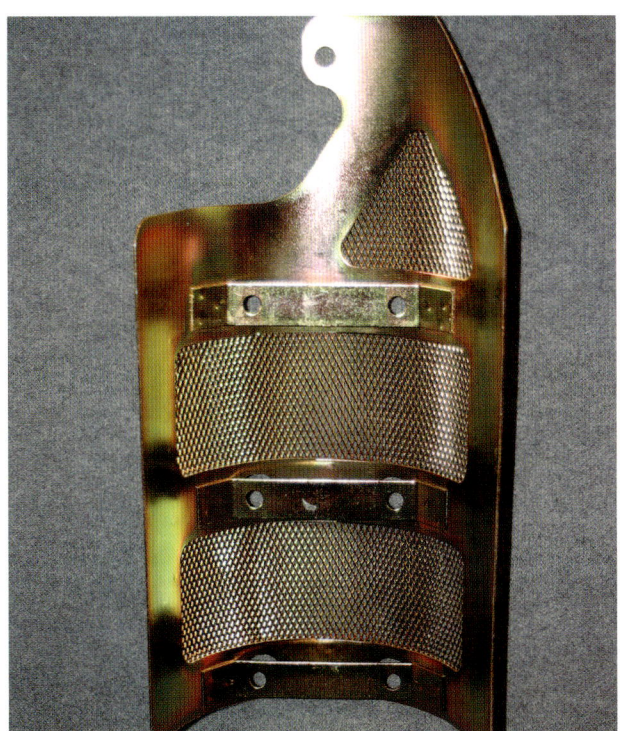
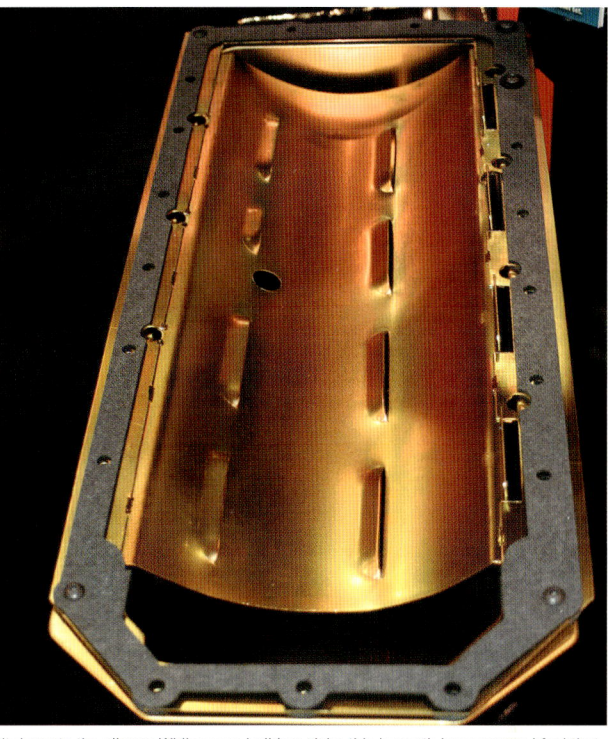

A windage tray captures oil from the spinning crankshaft and connecting rods and directs it down to the oil pan. While some builders claim this is worth horsepower, I feel that centrifugal force will be throwing the oil off the reciprocating assembly anyway, so power gains might not be possible. But getting precious oil back to the sump is a necessity, and the faster this happens, the better! Both screened and louvered types are available; both of these are Milodon products.

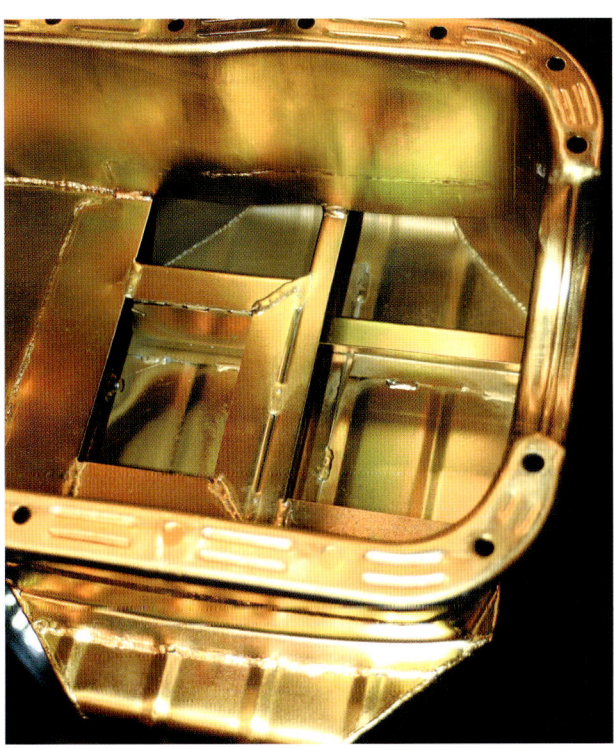

Oil control is a critical element in performance engines. Lubrication is life to an engine, and if it's being pushed hard on a regular basis, a factory oil pan may not direct lubricant to the pump pickup in hard launching or cornering situations. To cure this, one-way doors (or baffles) in the pan keep oil available to the pump. This is a Milodon road race pan, with a wider-than-stock lower sump, and trap doors that swing left and right to trap oil as the car turns. A drag race pan would have trap doors that faced toward the rear, and would hold oil in place as the car launched forward.

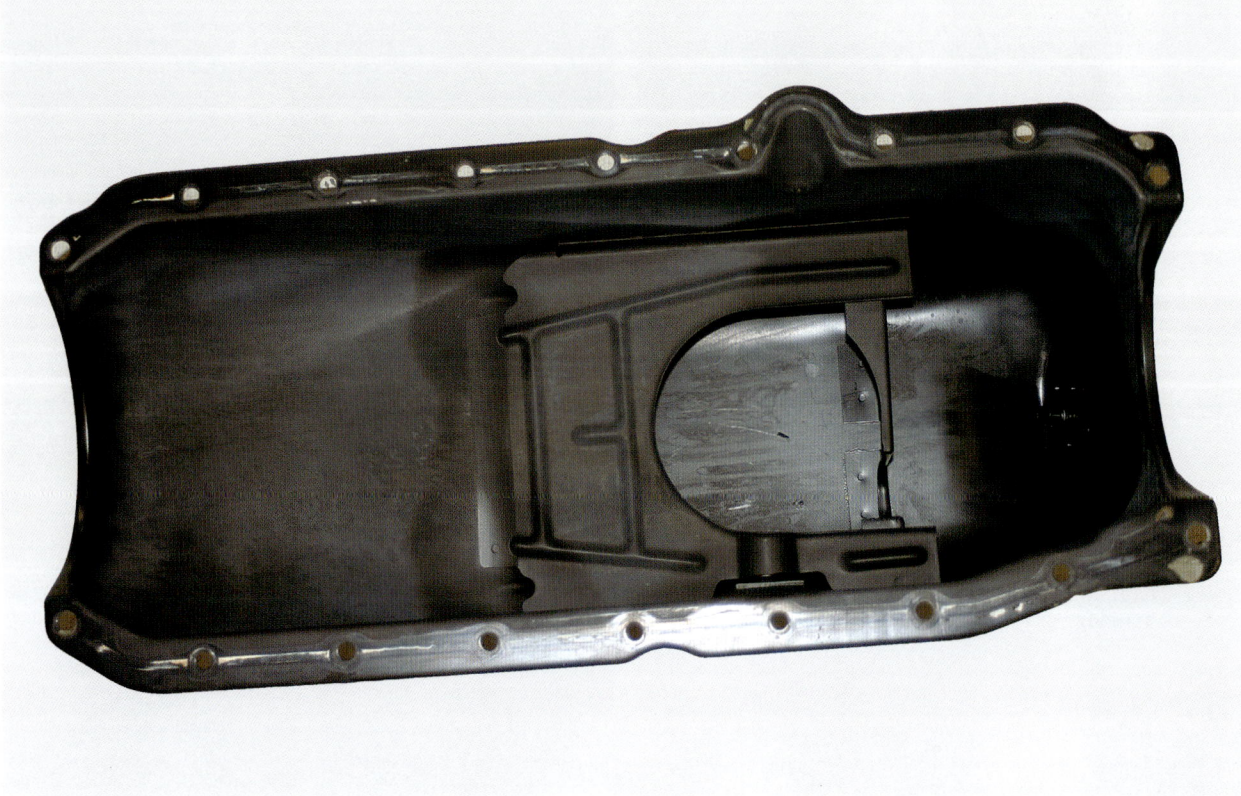

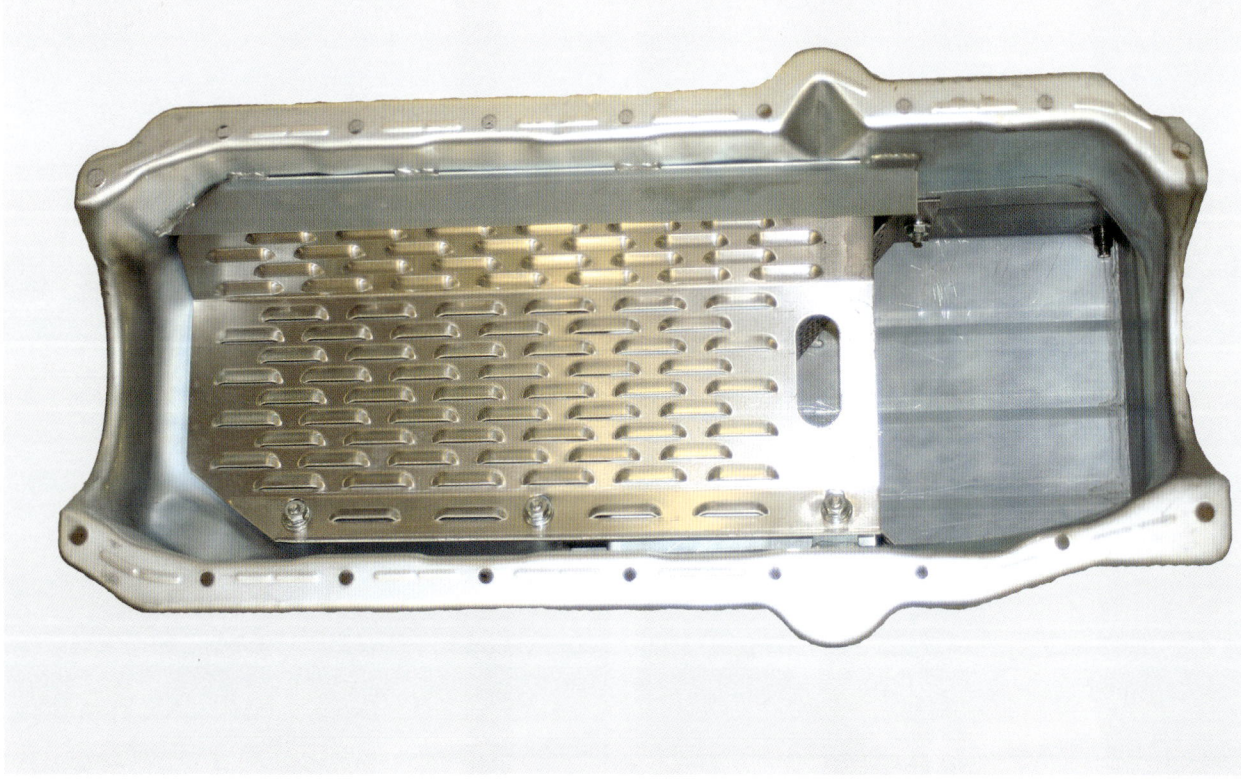

When we look inside a stock pan next to an aftermarket drag race pan, the differences in design become clear. The integral windage tray in the aftermarket pan is much larger to limit the amount of splashing in the oil pan as lubricant flies off the spinning crank and connecting rods. Also, the stock pan is stamped steel, while the aftermarket part is fabricated aluminum. This saves a little weight, but sheds much more heat.

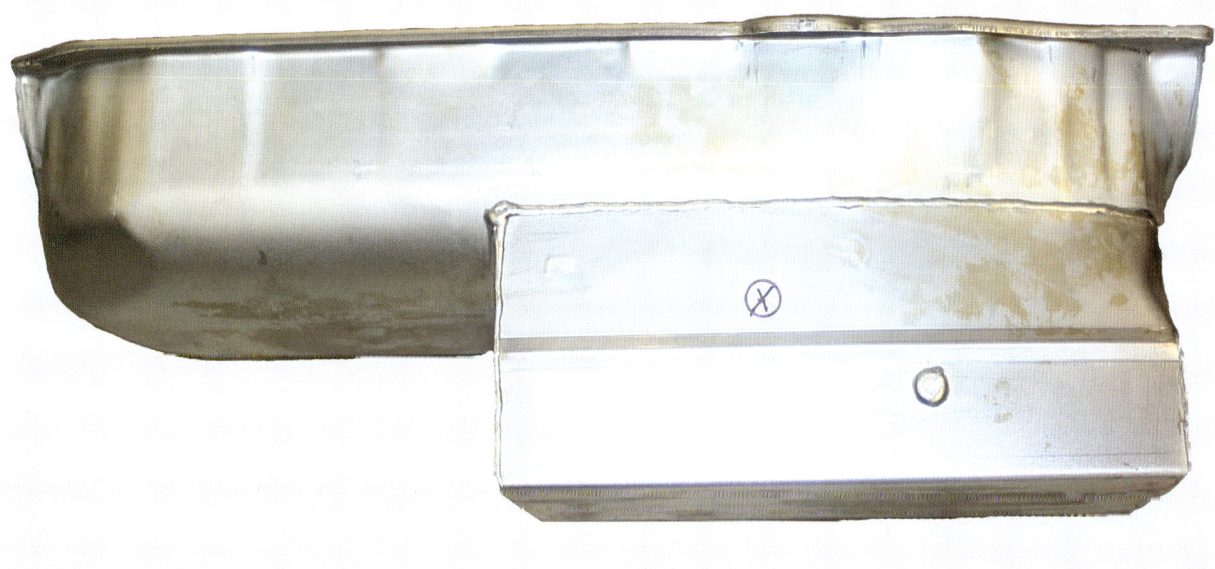

The difference in pan depth and width is obvious when we look at the profiles of each pan side-by-side. The added volume allows for an extra quart of oil in the aftermarket pan, so more lubricant is available to the engine during a high-speed acceleration run. A shallow stock pan could literally be sucked dry during a quarter-mile pass without adequate quantity and control of the oil inside it. Use of the aftermarket pan ensures the pickup will have oil to pump all the way down the track.

Some factory oil pump pickup designs are less than optimal, and some aftermarket oil pans require special types of pickups. The aftermarket has responded with well-designed oil pan pickups to provide efficient gathering of large quantities of oil. These are Milodon designs.

engine. In some race applications they can be justified, but these cases are rare indeed.

The oil pan is a simple sheet-metal stamping that has a responsibility and purpose beyond its humble appearance. The oil pan serves as a sump or reservoir for the engine's lubricant, it serves to cool the oil (as a heat sink), and it delivers oil to the pump pickup. Inside the oil pan, it's common to find baffling and/or trap doors to control where the oil goes while the car is moving. This becomes more important as the car moves faster or begins operating closer to the edge of control. More g-force input on the car means the oil is getting thrown in different directions inside the pan, and baffles (sometimes teamed with one-way trap doors) can control the oil and keep an adequate supply fed to the pump pickup.

In situations where keeping a steady supply of engine oil to the pump pickup becomes a challenge, it makes sense to look over the entire pressure lubrication system and make improvements or upgrades where necessary. One of the biggest improvements that can be made is to add a windage tray and a crank scraper. These simple additions are nonmoving parts that pull oil away from the spinning crankshaft and direct it back into the oil pan, where it belongs.

A windage tray does this by catching oil flying off of the crankshaft and connecting rods into a screen or louvered tray. Without a windage tray, oil would randomly fly around the crankcase, upsetting the sump oil and delaying oil drainback. A crank scraper is mounted very close to the crankshaft counterweights and serves to wipe excess oil off the crank as it passes by. This simultaneously removes the weight of the oil from the crank while quickly returning it to the oil pan. When both a windage tray and a crank scraper are employed, especially when teamed with a well-designed, baffled oil pan, engine oil can be controlled effectively and the pump pickup has a better chance of doing its critical job.

Most of the best oil pan, windage tray, and crank scraper designs can be found in the aftermarket. Manufacturers of these parts have researched and tested these components, usually in a competitive racing environment, and it pays to look into them. Even if you are determined to run a stock, factory oil pan, there may be some internal modifications you can make to help it do its job. You can weld in baffles, add trap doors, or get the stock pan coated to help it shed heat. Depending upon the purpose of your project engine (street use, drag racing, road racing, etc.), different modifications can be incorporated.

The oil pump pickup should be 0.500–0.750 inch from the floor of the oil pan. To measure this, use two rulers to check the pan depth, and then adjust the pickup to be the correct height. Don't forget to figure in the thickness of the oil pan gasket into your measurement. Once the pickup depth has been set, the pickup can be secured in the proper position.

Most oil pump pickups are designed to be a slip fit, and there's no easy way to install them unless you have a tool like this. The U-shaped end of the tool fits snugly against the pickup tube, and it can be gently hammered into place so it is fully seated against the pump housing.

Aftermarket oil pans were first developed for drag racers looking for increased oil capacity. This was to help ensure there'd always be enough oil for the pickup, especially during high-rpm use when the pump would move a lot of the oil out of the sump. Later research showed that effective oil control could alleviate many of the oil starvation problems racers were encountering on the track, and today's oil pan designs show the results of that research. Typical aftermarket pans today may still hold more oil than their factory counterparts, but effective oil control measures have become a priority in their design. It's easy to spot drag-specific and road–race-specific pans, as drag pans typically have deeper sumps and baffles designed to control oil being thrust rearward as the car launches. The deeper versions sacrifice ground clearance and are known to hang pretty low in the chassis. While this isn't a problem at the drag strip, it may become an issue during normal street driving on bumpy or pothole-infested roads.

On the other hand, oil pans designed for road racing are typically wider than the factory parts they replace, and their internal baffles are engineered to trap oil near the pickup when the car swings either left or right in hard cornering. Because road race cars are typically very low to the ground, any additional capacity in the design is engineered into "kickouts" extending from the sides. While the use of a road race–inspired oil pan probably won't affect ground clearance under your street car, the wider profile of such a pan might require additional clearance on either side of your engine and impact your suspension or exhaust system routing. Luckily, most aftermarket oil pan manufacturers also offer stock replacement oil pans with adequate clearance all around and improved internal baffling. Many of these same manufacturers also produce oil pump pickups designed for use with these same pans. A matched set of well-researched components will surely work better than cobbling together bits and pieces from different places.

Positioning the pickup the proper distance from the floor of the pan is essential. Too little clearance will restrict how much oil the pickup can draw in; too much clearance will cause starvation while there is still oil in the sump. It's generally accepted that a half-inch of clearance is suitable, and if you're within an eighth-inch of that, you should be fine. It's also important to check that the pickup is perfectly level to the pan floor, as a crooked pickup won't be as efficient and may cause problems of its own.

The pickup depth and clearance to the oil pan floor is normally measured by placing a piece of modeling clay on

The pen points to the slip-fit joint where the pickup tube meets the oil pump. It is common practice for this joint to be spot-welded to ensure it remains secure and at the proper adjusted height.

For those uncomfortable welding in this sensitive area, Pioneer Products has developed a bolt-on bracket to secure the pickup in place for some engine designs. This simple solution is highly recommended. The bracket shown is for small-block Chevy engines, and carries part number 839061

the bottom of the pump pickup, and then installing the oil pan in place with all the oil pan gaskets and scrapers to be used. It's becoming more common for the oil pan rail to be used as a mounting surface for other accessories (like the main bearing cap girdle mentioned earlier, and also for crankshaft oil scrapers), but stacking such things between the oil pan and engine block will alter the oil pump pickup depth. Once you have everything mocked up just the way it will run in the car, tighten the oil pan bolts to the specified torque value. When you remove the pan again, the modeling clay you placed on the pump pickup will have been crushed to the bottom of the pan. Measure its thickness to know exactly how much clearance you've got and if any further modifications are required.

Different domestic V-8 engine designs have widely varied oil pump pickup designs, so there's no universal means of adjusting the depth of the pickup. Be sure to research what your pump needs.

Once the depth has been set correctly, I prefer to weld the oil pickup system into place before the oil pan is installed for the last time. It is common for slip-fit and swaged components to work themselves loose, causing a loss of oil pressure and destruction of engine parts. This is not something can be easily seen or even checked, so it makes

sense to secure the oil pump pickup completely so it cannot become an issue. I weld every joint in the plumbing all the way to the oil pump itself. Welding the pickup tube directly to the pump might seem like overkill, but I like knowing it cannot work itself loose or even turn.

Similarly, the oil pump must be secured in place to the block. We've already talked about ensuring that the pump passageways are aligned properly and have adequate flow into and out of the block, so securing the oil pump in place is the next logical step. Some domestic V-8s rely on a single bolt or stud to secure the oil pump, and should this single fastener become even slightly loose, it will immediately affect oil pressure throughout the entire engine and probably cause some damage. You've probably heard stories about expensive engines being ruined by a $1 part, and the oil pump mounting hardware surely qualifies. I encourage the use of secured fasteners and liquid thread locker in this critical area, and have even safety-wired oil pump mount bolts together in race applications where there was no room for error. Inside a high-performance V-8 engine, the heat and pressure can be immense, and the range of temperature change is very wide. This is exactly the kind of environment where fasteners can work themselves loose, and a little overkill is just enough.

Chapter 3
The Crankshaft, Connecting Rods, and Pistons

CRANKSHAFT

The hardest-working part in your engine is probably the crankshaft. It is responsible for carrying more loads in more different directions than any other component. It must contain and transfer immense levels of power from a vertical direction (as the pistons and rods move up and down) to a horizontal twisting motion (as it spins into the transmission) and it must do so under heavy loads.

Add the draw of accessory drive belts on the front of the crankshaft and the transmission load on the rear of the crankshaft, and you'll soon begin to understand all of the different forces at work on this one critical component. It is being pushed, pulled, and twisted in many different directions at the same time, and these energies are amplified in a modified application. It pays to have a top-quality crankshaft when you're planning to throw a lot of power at it, and makes even more sense if more power might be introduced in the future. It's common for hot rodders to find more power once their engines are complete and installed for a while, and knowing

the crankshaft is capable of handling more is a good investment in the future.

The most important factor to consider when shopping for a crankshaft is the material it's made of, and how it was manufactured. The overwhelming majority of domestic factory V-8 crankshafts are either castings or forgings in 4340 steel. The castings are typically weaker than the forgings, and for high-performance use, a forging is preferred. The pressure exerted upon the crankshaft during the forging procedure makes for a tighter grain and greater strength when compared to a casting. The denser structure of the forging also adds some weight, but this is an acceptable compromise.

The best racing crankshafts are whittled out of a solid chunk of billet steel, and they are justifiably expensive. The next-best choice is a steel forging. Some of the best factory cranks and most aftermarket units are forged, and are engineered to support big power. There are different

Shown are both factory and aftermarket crankshaft rod journals. Note how sharp the transition is at the end of the journal on the stock crank (left), and how sharp the edges of the oil hole are. Compare these to the gently radiused transition and oil holes on the aftermarket unit. Cracks begin on sharp edges, so rounding off these edges prevents cracks from starting.

Rounding off the edges of a crankshaft counterweight helps in two ways. First, weight is removed from the crank, which helps it gain rpm at a faster rate. Secondly, the crank gains aerodynamic shape, which helps it cut through the thick, oily air inside the block. If any liquid oil splashes onto a counterweight, the aero shape helps shed it. Compare the aero shape of the aftermarket crank on the left to that of a factory crank, and it's easy to see the difference.

varieties of cast cranks, including (in order of preference) cast steel, nodular iron, malleable steel, and finally cast iron. Researching the crankshaft you intend to run and getting a professional opinion on it from your machinist is always a good idea.

Cranks can be modified to support more power through heat treating, nitriding, reradiusing the fillets, having the oil holes radiused, and polishing the bearing surfaces. These processes are all in addition to the standard machining processes like resizing, checking for straightness, and balancing. Lightening crankshafts is becoming popular as reshaping the counterweights to be more aerodynamic gains acceptance. The aerodynamic shape cuts through the air and oil film present in the crankcase with less resistance, and if some weight can be cut without upsetting the engine's balance or sacrificing strength in the crank itself, this is a good move. There are also some new oil-shedding coatings that can be applied to the nonmachined portions of the crankshaft. These coatings serve to keep oil off the crank (and connecting rods, if they are also coated); keeping the reciprocating assembly dry helps minimize the weight of the spinning mass.

The stroke of your crank is typically limited to what is available for your given block. As a hot rodder, I have always opted for the longest-possible stroke that would physically fit inside my block of choice. An engine with a longer-than-stock stroke is typically called a "stroker."

There is no question that developing an effective short-stroke high-performance engine is possible, and opting for a shorter crankshaft stroke is not always a bad thing. However, when you look at the basic horsepower formula, it's clear a larger-displacement engine simply has more power potential than a smaller-displacement version. A larger-displacement engine is a better choice for a street machine, as it will be capable of making more power at a lower rpm level than an identically equipped engine with a shorter stroke and less displacement.

A longer-stroke crankshaft will be heavier, and it will take more time to accelerate and decelerate than a shorter-stroke version. While some racers see this as a reason to run a shorter stroke, the odds are good that the longer-stroke crank will make more power throughout the acceleration curve than the comparable short-stroke version. You need to choose if you want an engine that accelerates quickly at a lower overall power level, or a more powerful engine that may rev slower. Personally, I'll take more power every time, unless it's in a racing application where having too much power on tap might challenge the traction capability of the car. These situations are rare, and so are the times when I'd prefer a shorter stroke. Not having to spin the engine

A typical stroker kit contains the crankshaft, connecting rods, and pistons as a matching set. Many are even offered as balanced assemblies, complete with bearings. The set shown is the popular 383-cubic-inch upgrade for the 350-cubic-inch SBC from Scat. The crank is clearly marked with the new stroke dimension (3.750), a jump from the factory 3.48-inch stroke.

With additional stroke comes reduced clearence. On this stock block (left), we see the adequate clearence to the rod bolt with the stock crankshaft, verses the interference we find with the stroker crank (right). Most factory blocks can be notched in this area to provide adequate clearance. Additionally, the head of the connecting rod bolt could also be ground down to clear the block if necessary.

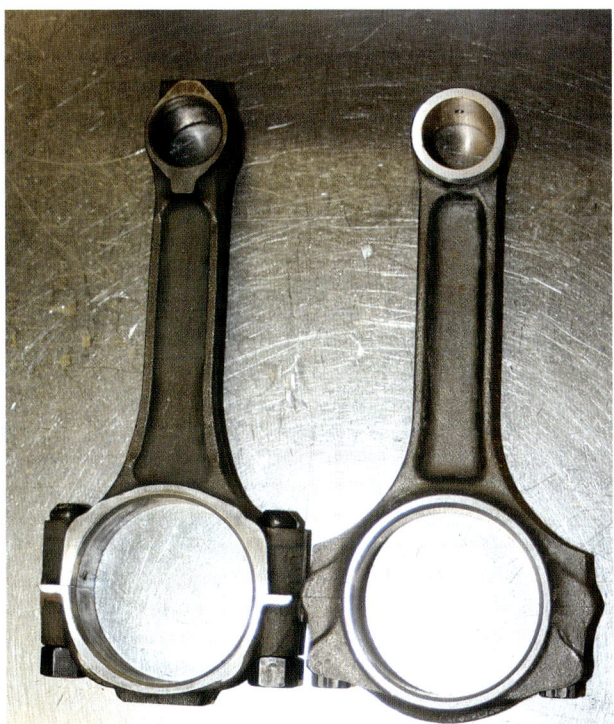

The Scat rods are designed for use in stroker engines, with additional clearance engineered into them at the critical shoulder area right above the rod cap bolts.

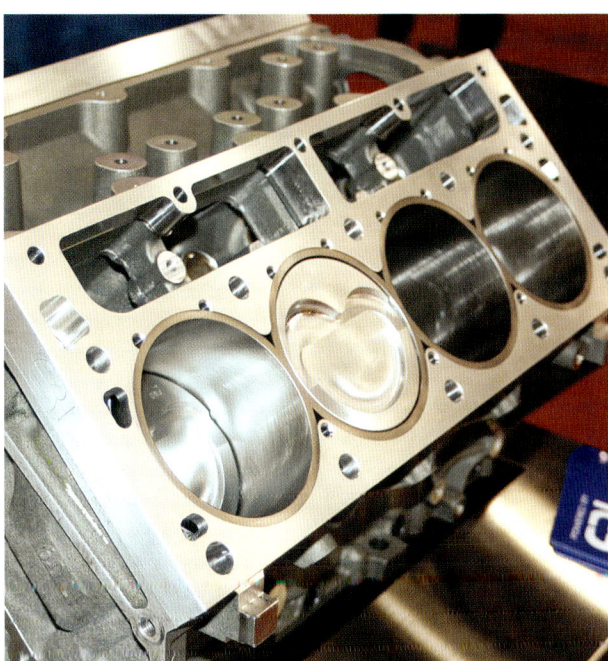

This LS1 small-block now displaces 500 cubic inches. The use of a spacer plate to make the cylinders taller (sleeves pass through the spacer plate and into the block without interruption) gives enough room for a longer-stroke crankshaft to do its job. Custom spacer plates for the intake manifold will also be required, now that the cylinder heads will be mounted farther up and farther away from each other. The relatively easy job of fabricating spacers is worth the additional cubic inches and power the big engine can now make. The expensive part includes the custom crankshaft and custom-length connecting rods this engine will require.

to high rpm levels is an additional benefit, as high-rpm operation increases internal engine stresses.

It is also possible to add a bit more stroke to almost any forged steel factory crankshaft by welding additional material to the rod journals and offset-grinding them to add more travel from top dead center to bottom dead center. Naturally, this would require the engine builder to research the block for clearance for this new, longer stroke, as the connecting rods will cut a broader path too. Additionally, custom pistons may be required, so having a crank stroked for more displacement can be a complex upgrade.

In some applications, it's possible to use a crankshaft from a larger-displacement version of the same basic engine in a smaller-displacement variant. The most common example of this is in the popular 383-inch small-block Chevy, where a 350-cubic-inch block is fitted with a crankshaft from a 400-inch engine. This upgrade has become popular because of the availability of good 350 blocks to build upon. While my mindset would suggest using a 400 block to get the most possible cubic inches, good, rebuildable 400-inch blocks are getting a bit tougher to find. Good 350 blocks are plentiful and cheap—even in the desirable four-bolt main bearing version. For those on a budget, these are major considerations. A 383-inch small-block outperforms a comparable 350 across the board, and its popularity has even prompted aftermarket crankshaft makers to offer complete upgrade kits. Based on its success, more and more stroker crank kits like this are now offered for a wide range of domestic V-8s. It makes good sense to research what's out there before investing any cash in your old stock crank.

Certainly, if a new stroker crankshaft is available in ready-to-run condition for the same or similar price as it would take to machine, balance, and upgrade a stock crank, it's worth consideration. As mentioned, the step up to a stroker crank typically means different connecting rods, and different pistons may also be required, so you really need to do the homework ahead of time.

When I was researching a 383 build, I found a complete performance upgrade package that included the longer-stroke 383 crank with matching rods and pistons. These were all balanced together (versus the externally balanced factory 400-inch Chevy cranks, which require unique externally balanced harmonic balancers and flywheels) and as such would save me the costs associated with balancing the reciprocating assembly. To purchase a good-quality 350 crankshaft, suitably strong connecting rods, and good-quality forged pistons, and then to have them balanced together, is a pricey proposition. The longer-stroke crank with the matching prebalanced rod, piston, and bearing package saved me money at the machine shop and allowed the use of internally balanced accessories, rather than more expensive, 400-specific accessories, balancer, or flywheel.

Engine	Stroke	Rod Length	Rod Ratio
Chevy 302	3.00 inches	5.70 inches	1.90:1
Chevy 327	3.25 inches	5.70 inches	1.75:1
Chevy 350	3.48 inches	5.70 inches	1.64:1
Chevy 350 w/ 6-inch rod	3.48 inches	6.00 inches	1.72:1
Chevy 400	3.75 inches	5.45 inches	1.45:1
Chevy 400 w/ 5.7-inch rod	3.75 inches	5.70 inches	1.52:1
Chevy 400 w/ 6-inch rod	3.75 inches	6.00 inches	1.60:1
Chevy 396/427	3.76 inches	6.135 inches	1.63:1
Chevy 454	4.00 inches	6.135 inches	1.53:1
Ford 289	2.875 inches	5.156 inches	1.79:1
Ford 302 (Windsor)	3.00 inches	5.09 inches	1.70:1
Ford 351 (Windsor)	3.50 inches	5.954 inches	1.70:1
Ford 460	3.85 inches	6.605 inches	1.72:1
Mopar 318/340	3.31 inches	6.123 inches	1.85:1
Mopar 360	3.58 inches	6.123 inches	1.71:1
Mopar 383/400	3.375 inches	6.358 inches	1.88:1
Mopar 413/426W/440	3.75 inches	6.768 inches	1.80:1

CONNECTING RODS

Once you've made a firm decision regarding your crankshaft, it's time to look into connecting rods. Deciding how long the rod should be and what material it should be made of comes next, and there are some basic considerations to look at here.

There are many opinions on rod length and rod ratio (the ratio of the stroke length to the rod length). For many years, performance-engine builders generally agreed that a longer rod was preferred. This lessens side loads. (On the power stroke, the pressure forcing the piston downward acts upon the rod's angle, and uneven pressure against the cylinder wall results.) A shorter rod will have a greater angle, and therefore will produce a greater side load than a comparable, longer connecting rod. Also, a longer connecting rod will spend more time at both top dead center (TDC) and bottom dead center (BDC) as it changes direction. This extended dwell time is seen as an advantage because of the time it takes for fuel to burn. While the fuel is burning and pressure is increasing, the added time the piston spends at TDC can allow for more potential power transfer. These theories are offset by the fact that longer connecting rods weigh more than comparable shorter rods. Fans of shorter rods also argue about exactly how much benefit is truly seen at the flywheel (or at the drive wheels) at the lower rpm ranges.

Rod ratio can be calculated simply by dividing the rod length by the crankshaft's stroke length (rod ratio = rod length / stroke length). If the engine has great breathing potential, a rod ratio of 1.75:1 has traditionally been considered the ideal.

Personally, I feel connecting rod length should be determined by the piston. As we'll soon discuss, many different piston configurations are available, and a well-designed piston has a dramatic impact on power potential and longevity. If you choose the best-possible piston for your application, your connecting rod length will have to work with this piston. Where the piston pin is placed in your piston of choice is one of the critical choices you'll have to make, and once decided, this clarifies your rod length options.

I typically try to find common connecting rods (or even rods with common big- and small-end sizes) that will work for a given application. If connecting rods designed for use in a small-block Chevy will work in your non-Chevy engine build, you can save a lot of money you'd otherwise spend on custom rods. Naturally, you'd have to get the crankshaft rod journals machined to Chevy dimensions, but this is becoming a common request. Similarly, using common Chevrolet piston pin diameters is also gaining popularity, as aftermarket pistons (and piston pins) are less expensive for these common engines.

The rods have a tough job to do. They are stressed in opposing directions at varying levels of force with every turn of the crank, and this is not a job to be taken lightly. It is not unusual (or even unexpected) for a connecting rod to break in a

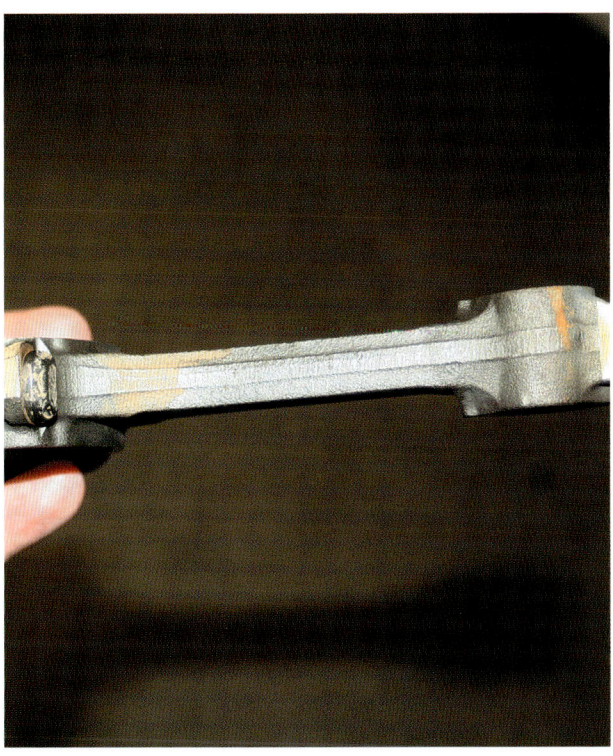

Stock connecting rods can gain a bit of durability and strength and lose a bit of weight by grinding down the parting line along their beam. The parting line is pronounced, and you only need to remove enought material to make it smooth.

A belt sander makes the job much easier, as it can remove most of the material in just a few minutes.

Once the rod beam is smooth, you can move to the bench vise to finish up the job.

Remove the remaining material along the parting line with a cartridge roll until the entire beam is polished.

performance engine. The more power the engine makes, or the more rpm it's pushed to, the more likely it is for a connecting rod to give out.

This is not to say you're guaranteed to break a rod if you build a performance engine. If you choose connecting rods with plenty of strength, you might never have to worry about them. Luckily, there are plenty of choices out there, but the premium options carry a high price tag.

Most street-engine enthusiasts are fine with having their stock connecting rods inspected, resized, and upgrading the rod bolts to premium hardware. The piston pins may be either pressed into place (where they fit snugly in the connecting rod, and only provide movement on the piston) or floating (where the pins are able to rotate inside the connecting rod as well as the piston). If you're stepping up to full-floating pistons, the connecting rods should be bushed as well. The bushing (typically made of bronze) serves as a bearing surface for the rotating piston pin.

Some enthusiasts go a step further with the stock rod prep and grind down the beams of the rods to eliminate any potential stress risers and to shave a bit of weight. Naturally, all of this work should be accomplished before balancing the reciprocating assembly.

For those who will demand more power from their engine than stock connecting rods can handle, the aftermarket is ready to help with a wide range of options. There are stock-type I-beam-design rods made of top-quality steel, which are typically manufactured to high-quality standards and ship with the desired high-performance hardware already installed. These rods are great for street engines that see occasional track time, as they are fairly priced and offered as stock replacement units for popular engines. The next step up is an H-beam steel rod, which is typically stronger than a comparable I-beam design. They are also wider, due to the design, and may cause fitment issues in some engines. The H-beam design may also be a bit heavier than comparable I-beam units, but if their strength is necessary for the engine to live, this is one sacrifice that will have to be accepted. Aftermarket H-beam connecting rods always ship with top-notch hardware, and may even have been designed for the connecting rod bolt to thread directly into the rod itself (as opposed to using a bolt and nut) for additional security.

Quality forged aftermarket I-beam and H-beam connecting rods are available in a wide range of lengths and end diameters. If you're screwing together an oddball combination that will require a special rod, you should know that there are people out there who will make the rod you need, but it will not be inexpensive.

While the overwhelming majority of enthusiasts will never need more than a steel rod, aluminum connecting rods are available. These are typically designed for drag racing engines that will not run for long periods, but that will see immense pressures when they do. Aluminum rods are lightweight, and some enthusiasts have run them on the street for many miles, but I'd never recommend this for a true street machine. Aluminum rods are typically larger than comparable steel rods, so clearance may be a concern. Also, aluminum rods will stretch more than steel at high rpm levels. Compensating with a bit more piston-to-valve and piston-to-deck clearance is the key when using aluminum rods. Fatigue in is also a concern, but because aluminum rods are used almost exclusively in regularly

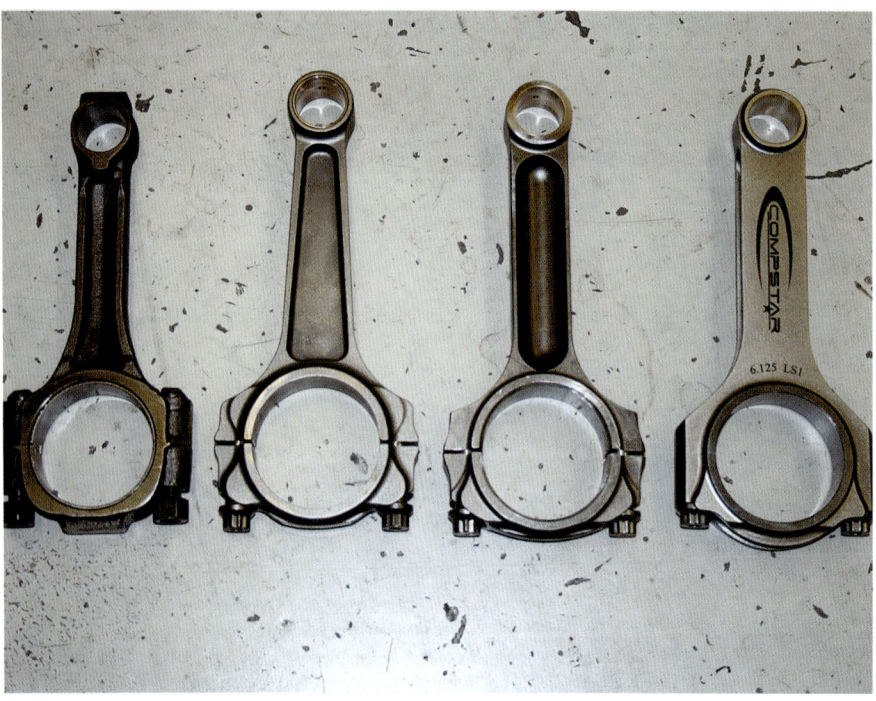

If you're going to step up to a better-quality connecting rod to replace the stock units, many great options exist at all budget levels. Here are four connecting rods, ranging from a cast factory rod on the left, to a pair of forged aftermarket I-beam rods in the center, and a race-level forged H-beam rod on the right. Of particular note is the hardware used to secure the rod cap to the beam. The factory hardware cannot compare to the top-notch aftermarket bolts. These bolts are critical for strength and durability.

Connecting rods rely heavily on the rod cap bolts. While it's easy to understand that a larger-diameter bolt can provide increased security and clamping force, it's not simply a matter of installing larger bolts in the existing rods. Look how much more material this aftermarket rod has around the fastener (right) when compared to the factory rod it replaces (left). This material is in a critical area, and the strength here is necessary to support the larger-diameter, high-strength fasteners.

rebuilt racing engines, builders have a chance to inspect them often. Street enthusiasts don't have the chance to check out their engine's internals on a regular basis, so a good set of well-prepped steel rods is the way to go. In a high-rpm, full-race application, aluminum rods would be preferred.

The typical domestic factory V-8 built since 2000 will probably have powdered metal (PM) connecting rods. These are an interesting bit of technology, and can support a fair amount of power in most cases. They are made by pouring powdered metal into a mold, and then heating and compressing it under high pressure into a precise shape. This allows for very accurate control of the rod's sizing, and the consistency of the process creates very little weight deviation from rod to rod. The manufacturing process produces a smooth surface on the rod, so there's no need to grind or polish the rod's beams to eliminate stress risers. For some, the cheap and plentiful PM rod is a good choice. For example, GM's PM V-8 rods are lightweight, inexpensive, and can support 400 horses. If a 400-horse engine is your goal, PM rods might be a great option for you.

Performance enthusiasts have reservations about PM rods because of their power limitations and because they're so inexpensive. Additionally, the rod caps are cracked off during the manufacturing process, which doesn't appear very precise or impressive to custom engine builders who are used to the fine quality of aftermarket rods. The cracked caps are all unique to the rods they're made with and must be kept as a matched set, but the complexity of the break also means the caps are less likely to ever walk or shift around. While they'll never be racing parts, stock PM rods with upgraded hardware are a perfectly acceptable choice for milder engines, and their availability and

Compare the end of the I-beam rod to the H-beam unit. The H-beam rod has no discernable weight pad on either end, and is made to a high standard of quality to be identical in weight to the others in its set. The higher level of quality control in its manufacture results in both a lighter weight overall (due to the lack of a weight pad) and a higher cost. Also note the rod end is drilled for improved piston pin oiling—another benefit.

This is a stock GM powdered metal (PM) connecting rod with a cracked cap. As part of the manufacturing process, the rod cap is broken off to separate it from the rest of the connecting rod. This actually works quite well, and the precise fit between the broken halves minimizes the chance for the rod cap to move or "walk." The grooves along the parting line on the inside of the bore are part of the original one-piece precision casting, and when the cap is snapped off, they offer a natural weak point for the break to happen.

affordability should keep them on the lists of budget-minded street engine builders.

Naturally, the length of the connecting rod you choose is based on your piston choice. While some choose to run a the longest-possible rod and the shortest-possible piston, recent research into sub-6,500-rpm engines has shown precious little difference in the power production of similar engines with differing-length rods and pistons and no other changes. The most common example includes the most common engine choice: the small-block Chevy. In a typical 350, the rod length is 5.7 inches. A common upgrade has been to run a 6-inch-length rod and a shorter piston, with the rings located higher up on the piston and with a shorter piston skirt. The differences in power are so minor that the additional cost of the custom piston/rod combo really can't be justified. In a racing application, where horsepower is critical and the engine sees rpm levels past 7,000, the difference can be justified, but for the majority of street enthusiasts, it cannot. Taking advantage of the broad range of stock replacement pistons engineered for use with stock-length rods is a wiser move. Even if you want to upgrade to a stronger aftermarket connecting rod, those duplicating the stock length are usually the least expensive, because they are made in the greatest quantities.

When designing a custom engine, rod length is not a major concern of mine. More important is selecting the proper piston for the application; the connecting rod length will be what it needs to be to place the piston in the best-possible position at TDC. After a block has been decked and the cylinder heads have been resurfaced, the position of the piston relative to the combustion chamber will be altered. The thickness of the head

Balancing is the procedure where the weights of the connecting rod and piston are equalized (by the removal of weight from the heaviest members of a set to match the lightest) and then the crankshaft counterweights are fine-tuned to match. The crankshaft shown here has bob weights bolted to it that simulate the weight of the chosen connecting rod and piston assemblies. The assembly is spun on this machine and any out-of-balance counterweights are modified (either through drilling to remove weight, or the addition of heavier-weight metal to add weight) until the assembly is perfectly balanced.

Here's a good example of a quality aftermarket piston. This one is from Mahle. It is a precision-crafted piece with two different coatings, including a dry phosphate lubricant to protect against ring microwelding and pin galling at initial startup and Mahle's proprietary Grafal anti-friction skirt coating. Mahle pistons have CNC-finished pin bores, so no honing is required. These pistons come complete with high-quality steel pins, round wire locks, and low-drag performance rings, and represent a great choices available in today's aftermarket for what V-8 enthusiasts typically build—a performance engine requiring stock replacement-type pistons.

gasket (when compressed) will also factor in, and these variables must be considered and researched before the true target rod length can be determined.

The weight and balance of the reciprocating assembly has been mentioned before, but it's important to mention it once again. An engine builder should be able to mock up the entire assembly and determine that everything is perfect before final balancing takes place. If changes need to be made (such as a minor grind of the rod bolt head for clearance, or a change in piston height to get the proper compression ratio), all of these issues must be addressed prior to the final balancing of the reciprocating assembly. It would not be good to have to rebalance the engine because of an overlooked issue. Minimizing the surprises during an engine build saves both time and money, and frequent checks and double checks on clearance and fit help with this.

Balancing is accomplished by a trained pro, and is not a procedure an amateur can perform at home. If the reciprocating assembly is balanced to within a gram, the engine will reach higher rpm with smoothness, which is our goal. Some have questioned the need for balancing, but as someone who has worked on many custom engines, I don't question it a single bit. When teaming parts built for extremely heavy work, and combining components manufactured by different companies, the balancing procedure ensures the final grouping is fully compatible. The relatively minor investment in engine balancing

is easily justified by considering the result of a lack of balance: a shaky engine that wears out bearings prematurely.

To summarize, the connecting rods are of critical importance, and will require some investment to survive in the high-stress environs of your engine block. They should be as strong as possible while being as light as possible, and high-strength hardware is a necessity. Rod lengths are best determined by the components around them, and balancing of the reciprocating assembly must be accomplished once a firm mockup of all the relevant components confirms that their design and compatibility are sound.

PISTONS

The piston of a high-performance V-8 has a lot of work to do. Its job is critical, and the stresses it sees as it contains and (hopefully) enhances the power of the engine are quite high. It must be able to withstand these stresses long-term, through a wide range of heat cycles. That's a lot to ask, but the latest piston designs are better than ever. The advent of computer manufacturing (through computer numerical control, or CNC) has made custom pistons easier and less expensive to manufacture quickly.

Every facet of the piston's design, from the basic diameter to the height and spacing of the rings, to the positioning of the piston pin, to the length of the piston's skirt, is variable. Of critical importance are the design and shape of the piston's

Compare the Mahle piston on the left to this custom-crafted piston from Ross. I had this piston custom-made for a special 601-cubic-inch Pontiac engine. Notice how the shape of the dish reflects the shape of the combustion chamber for the best-possible burn while retaining a good quench area around it. The 601-cubic-inch Pontiac ended up making 740 peak horsepower on pump gas, thanks to these pistons and a lot of help from Butler Performance.

When we put the two pistons side-by-side, the differences become more obvious. The top ring in the Mahle piston is close to the top of the piston deck, which helps efficiency and burn quality by keeping air/fuel mixture closer to the spark plug. The short skirt and antifriction coating can also be seen. The custom Ross piston has a much thicker deck, which forced the ring to be moved down, but it will withstand the nitrous hit this engine will see. The Ross piston also has a much more substantial skirt and thicker rings, which will add weight but also give needed stability to the piston in this 1,100-plus-horsepower application.

top, or crown. The newest high-efficiency piston tops are a mirror image of the combustion chamber. This creates a large squish (or quench) area, where the flat portion of the piston top (basically, any area not covered by the combustion chamber) will come very close to the head when the piston is at TDC. It's not unusual for the piston in a well-designed street application to come within 0.030-inch of the head. This forces the air/fuel mixture into the combustion chamber, resulting in increased efficiency and greater power.

The farther away the air/fuel mixture is from the ignition source (the spark plug), the longer it takes for the mix to burn. Additionally, the mix that is farthest from the plug is typically the dirtiest, since it may not burn completely. The intricate edges and voids present between the top of the highest piston ring and the top of the piston offer the least efficient burn, and minimizing these areas with a well-designed piston pays off in many ways. Consider the dramatic difference between a piston with a fully dished top facing the combustion chamber versus a piston with a nice quench area and the chamber's shape reflected upon it. The dished piston will allow a relatively large percentage of the air and fuel to burn completely out of the combustion chamber, and it will not burn as well because the flame has so far to travel to get to it. If the combustion process is limited to the chamber, its shape can work with the air/fuel mixture to burn quickly and completely, requiring less ignition advance and resulting in more power across the board.

No single component influences the engine's compression ratio more than the piston. Typically, engine builders will base the design of the piston on the design (and volume) of their cylinder head's combustion chamber. If the fuel you intend to use will be a limiting factor (like pump gas for the street), getting the right compression ratio is of

critical importance. The compression ratio also adds to the efficiency of the burn, because a mix under high pressure will burn more effectively. If the fuel could stand it, we'd all be running higher-efficiency, high-compression engines. Since typical pump gasoline has low resistance to preignition and detonation, the compression ratio of a street-bound engine must be limited. Advances in the development of combustion chamber, camshaft, and piston design technologies have helped this situation greatly, however.

Determining your compression ratio is a critical and integral part of engine design. The compression ratio is a comparison of the volume within each of the engine's cylinders when the piston is at BDC versus when it is at TDC. The difference in these volumes is expressed as a ratio, and it reflects the amount the engine is compressing the air/fuel mixture prior to ignition. With some basic measurements, you can determine the compression ratio in any engine.

To determine your compression ratio, you must know the volumes of your cylinder, your combustion chamber, and your compressed head gasket. The cylinder volume can be found using the formula pi/4 x bore x bore x stroke. On a 350 Chevy with a 4-inch bore and a 3.48-inch stroke, the formula will look like this:

0.7853982 x 4 x 4 x 3.48

The total is 43.730971776 inches.

We've measured the cylinder volume in cubic inches, but graduated burettes (like the one we'll use to determine chamber volume) are typically marked in cubic centimeters (cc). So we must convert our cubic inch dimensions over to cc. This can

Here's a piston from a NASCAR Nextel Cup engine (Kasey Kahne's Dodge, to be precise). It's incredibly small height is an extreme effort to remove as much weight as possible. The ultrashort skirt offers little stability, but it only must live at sustained high rpm for a limited time.

be done by multiplying the inch dimensions by 2.54. In the case of our 350 Chevy, the bore dimension (4 x 2.54) becomes 10.16, and the stroke (3.48 x 2.54) becomes 8.8392. The new formula now looks like this:

0.7853982 x 10.16 x 10.16 x 8.8392

The total volume of the cylinder in cc is 716.622.

Determining the volume of the combustion chamber requires measurement with a graduated burette. The cylinder head, with valves and a spark plug installed, is mounted chamber side up. Then, a flat plate (usually made of plexiglass) is sealed to the bottom of the head. The plate will have a hole drilled through it to allow fluid to fill the combustion chamber. The graduated burette is filled with fluid (usually light oil or parts cleaning solvent) and then the chamber is filled through the hole in the plexiglass. Once full, the volume of the combustion chamber can be calculated based on burette readings.

Calculating the volume for the compressed head gasket is similar to determining the volume for the cylinder, since it's basically a much shorter version of a cylinder. The formula is now pi/4 x bore x bore x gasket thickness. For our theoretical 350 Chevy, let's assume the compressed gasket thickness is 0.045 inch. The equation becomes 0.7853982 x 4 x 4 x 0.045, which equals 0.565486704 inch. Again, we must convert to

cc, so the new formula becomes 0.7853982 x 10.16 x 10.16 x 0.1143. This translates to 9.266cc for the gasket volume.

We then add the cylinder volume to the gasket volume (716.622 + 9.266) we get 725.888cc. A typical chamber size for a high-performance small-block Chevy head is 64cc, so we'll use this dimension in our example.

Now we can add the volume of the cylinder, gasket, and chamber together to get a total volume of 789.888. This is the total volume with the piston at BDC. We divide this by the volume of the piston at TDC, which is the chamber volume (64cc) plus the gasket volume in cc (9.266). The total is 73.266.

The total volume is divided by the chamber-plus-gasket volume to get our compression ratio.

789.888 divided by 73.266 equals 10.781. This engine would have a 10.78:1 compression ratio.

Where pump premium gasoline–powered engines were limited to 9.5:1 compression years ago, today enthusiasts can enjoy 10:1 or even 10.5:1 with an efficient chamber and piston combination. The 9.5:1 compression ratio that enthusiasts looked at as the limit for 92- or 93-octane a decade ago can now be used in engines using cheaper 87-octane if the chamber is capable of staying out of detonation.

Other technologies are helping the piston makers today. Many ceramic-based coatings exist to insulate the combustion chamber, helping to deal more effectively with preignition

55

When placed next to a typical aftermarket SBC performance piston, it becomes obvious where the NASCAR piston saves weight. The support area around the pin is much more pronounced in the SBC piston, while the NASCAR part relies on a smaller-diameter pin with much less mass.

A side-by-side view makes the pin diameter difference more obvious, and the ultrathin piston rings, spaced closely together, minimize both weight and friction on the NASCAR piston on the right. The heavier, thicker rings on the SBC piston are designed to last long-term.

In the past, pop-up pistons were used to increase compression and make more power. Pop-up pistons were also known to block the flame travel, so they fell out of favor. This particular piston is for a Chrysler Hemi, so it fits into a spherically shaped combustion chamber and its pop-up design isn't a performance detriment. Note the dual dimples atop the piston dome—they allow the piston to clear dual spark plugs.

and detonation. When these coatings are used on both the chambers and the piston tops, and even the valves and ports, the result allows for higher compression and less detonation. I designed my 383-cubic-inch small-block Chevrolet to run on 87-octane, and coatings were part of the plan from the start. The resulting compression ratio was 9.7:1, and the engine made 545 peak horses at 6,400 rpm—on 87-octane gas! Credit for this stunning power also goes to the Comp Cams solid roller camshaft and CNC-finished Air Flow Research (AFR) aluminum cylinder heads, but the additional compression ratio was surely a contributing factor as well. I've not heard of other engines built to maximize power output 87-octane, but I'm sure they're out there now (or will be soon). The steadily rising cost of gasoline will contribute greatly to this, and planning for a higher compression ratio should begin long before you order pistons. Careful discussion with your piston maker of choice is the key, and working with them to get the most out of your pistons (and therefore your engine) is well worth the time.

Pistons are manufactured from aluminum, but how the material is shaped into pistons makes a big difference in how much stress the material can handle. Pistons can be cast aluminum, forged aluminum, or hypereutectic (infused with high-silicon aluminum), and these all contribute to the characteristics of the final product.

A forged piston will typically be heavier, but it will be capable of supporting much more power than a comparable cast unit. If making big power is your goal, a quality forging is the best choice. Forged pistons expand more than similar castings, and their assembly clearances are greater for this reason. Once they've reached operating temperature and the clearances are correct, a forged piston can be pushed hard without relenting. If you're considering a supercharger or nitrous oxide injection, it's hard not to recommend the excellent strength a forging will provide. Even when building an engine for a heavy car, I recommend forged pistons, because of the work they'll be asked to do. It's common for hot rodders to not take their vehicle weight into consideration when choosing engine parts. Surely, the entire build should be considered when designing any portion of the drivetrain, but with regard to pistons, I always make vehicle weight a priority. A typical V-8 will see much different stresses when asked to move a heavy vehicle (3,000 pounds or more) when compared to a lighter vehicle (under 3,000 pounds). Moving 3,500 pounds at wide-open throttle (WOT) is a much different experience for the engine than moving 2,500 pounds. The workload is greater, and the engine will typically be asked to spend more time at WOT to get up to an equivalent speed than an identical engine moving a lighter car or truck. This additional need justifies better-quality components.

In lighter-weight vehicles, a good-quality cast piston can often serve the need. Because cast pistons don't expand as much as their forged counterparts, their assembly clearances are tighter, and they have less of a need to warm up before

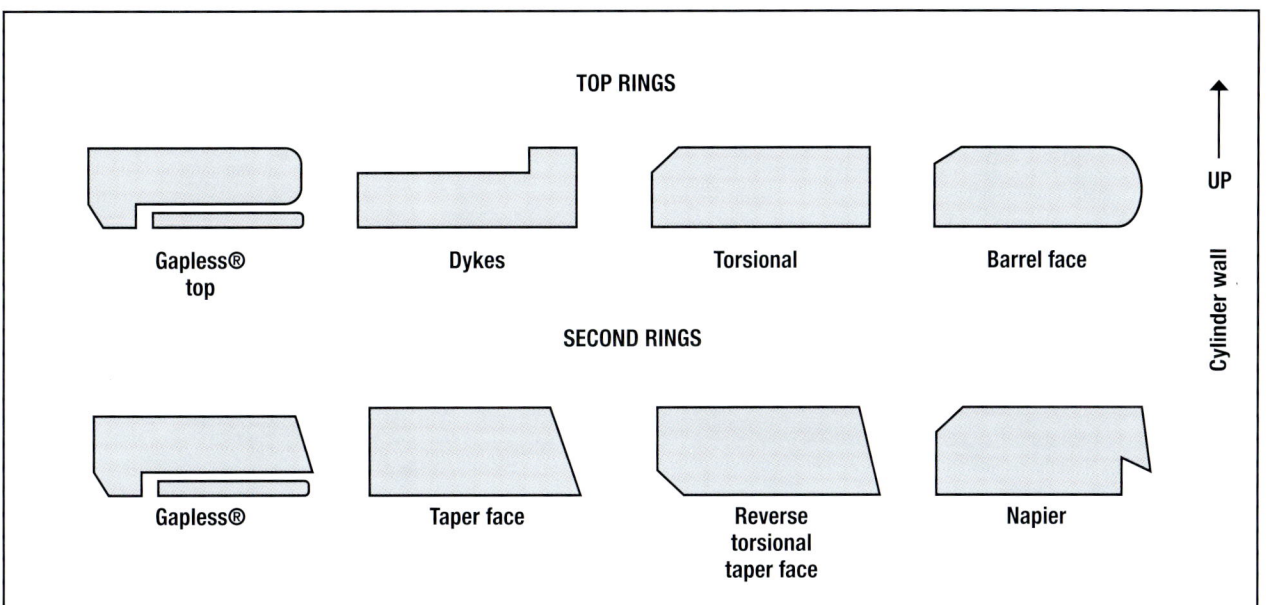

Above and below: Piston rings are very complicated bits of engineering that do a simple job—sealing the piston to the cylinder walls. The extreme heat and constant friction piston rings see has resulted in a wide range of products to work in different types of engines. *Images courtesy of Total Seal*

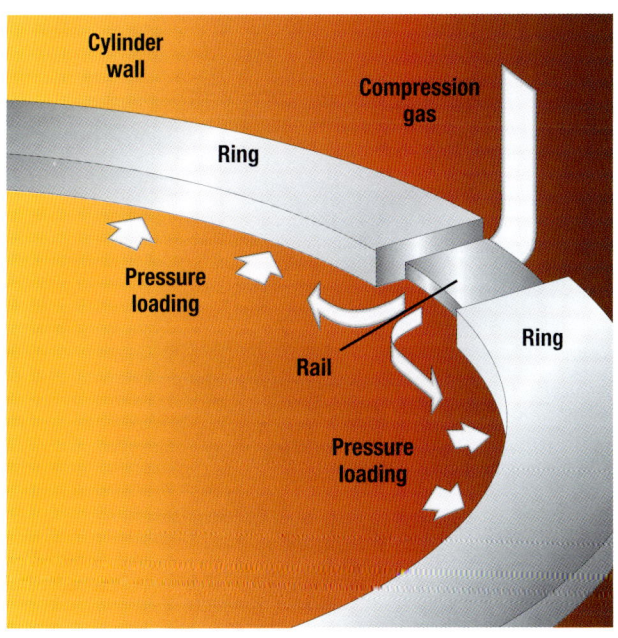

The advent of hypereutectic high-silicon content pistons opened a transitional choice somewhere between castings and forgings. They are manufactured using a casting technique, but additional silicon is added (between 9 and 16 percent). This makes the piston both harder and more brittle. The silicon crystals act as insulators, keeping heat in the chamber, where we want it. The silicon also limits the thermal expansion of the aluminum (allowing for even tighter assembly clearances), minimizing scuff wear and seizure potential.

If you're building a good street performance engine, oval track "claimer," or drag strip bracket engine, maybe a hypereutectic piston is a good choice for you. They are stronger than typical cast pistons, but they do have limits. They are typically heat-treated for additional strength and can support more power than a standard casting can. But if you intend to turn up the wick on power in the future, hypereutectic pistons might reach their limits. The reasonable price tag they carry makes them the perfect choice for some.

Piston design is paramount to performance, as every part of the piston offers choices. From the crown design on the piston's top to the skirts along the sides, many different options exist. The design of the cylinder head's combustion chamber should help you define the design of the crown and determine whether a dome, flat-top, or dish is right for you. Naturally, this is the direct result of your target compression ratio, and other factors (such as the swept volume of the cylinder and thickness of the compressed head gasket) must be known before final piston dimensions can be decided upon. For more common

they can be pushed hard. If a supercharger or nitrous oxide enhancement is not in your future, you don't intend to push the engine past 6,500 rpm, and your vehicle doesn't weigh 3,000 pounds, a cast piston may be perfectly suited to your needs. The reduced cost of cast pistons is tempting, and the quality and consistency of cast pistons has never been better than it is today. Cast pistons do have a place in the performance aftermarket, but like so many other choices, it simply depends on the big picture and overall design goals of the project.

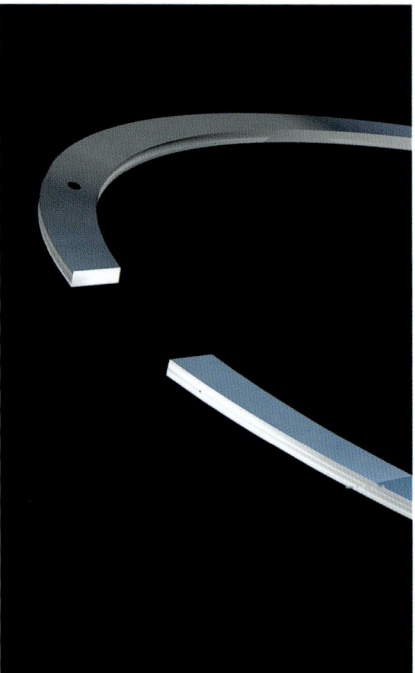

In addition to refining the performance of traditional (gapped) piston rings, Total Seal has been working with its own gapless ring designs. By experimenting with different piston ring materials and a wide range of plating and coating materials, piston ring manufacturers can supply products suitable for any application. Which is best for you? I recommend you ask them, as some of their new technologies may be perfect for your engine project. *Images courtesy of Total Seal*

engines, much of the homework and research has already been accomplished, and your piston manufacturer will probably offer several choices. For best efficiency and power, builders have learned that a shallow combustion chamber in the head, teamed with a well-defined dish shape in the piston, works best. This is because of the shape of the burn itself, which starts with the flame kernel ignited by the spark plug and spreads outward.

The huge, deep combustion chambers long offered by the factories and favored by performance engine builders were found to limit the efficiency of the burn sequence. To attain the high-compression ratios required to make big power, builders had to use high-dome pistons with these chambers. Commonly referred to as "pop-up" pistons, these high-dome designs extend past the top of the deck of the engine block when the piston is at TDC. These high domes were found to limit the flame travel across the chamber and piston top, costing efficiency and requiring more ignition advance to acquire peak cylinder pressure (and thus power) at TDC. The solution was to make the chamber shallower and offer a well-designed dish in the piston top. This method does not restrict flame travel, and research showed that a shallower combustion chamber would also deliver improved intake and exhaust flow, because the valves would no longer be shrouded by the deep walls of the combustion chamber. Now, the compression ratio can be adjusted by the depth of the dish in the piston rather than by a tall dome, giving more power from the same compression ratio.

The piston must be designed with enough material to incorporate the dish while maintaining sufficient deck thickness to deal with the power being made. Whittling a dish into a piston makes the piston's deck surface thinner, and this critical area cannot be compromised. This is especially true in supercharged, nitrous-injected, high-compression, and racing powerplants, where cylinder pressures and temperatures are pushed to very high levels. For this reason, I prefer to work with my piston manufacturer to choose a piston with a thicker deck surface that is capable of being dished without sacrificing integrity. It pays to work with the folks making your pistons to ensure they are a good fit with the rest of your engine package, especially the combustion chamber.

After the crown, the next logical thing to look into is the ring grooves. Their distance down from the crown is of particular importance. A top ring placed closer to the piston's deck will add some efficiency by keeping more of the air/fuel mix in closer proximity to the spark plug (and therefore the burn), but it will be exposed to more extreme heat and pressure as a result. The key is to have the ring down only as far as it needs to be, and this differs based on many different factors. A high-compression, supercharged, or nitrous oxide–injected engine will experience the high cylinder pressures typical of high-powered engines, and this will require a different ring package than a low-compression street mill. A compromise must be made, and relying on the experience of the piston maker is usually the best bet.

The oil control ring doesn't get the attention the compression rings do, but that doesn't mean design efforts have stopped. This new oil control ring from Total Seal was designed to break in easily, last a long time, and be compatible with the new sleeves being used in aluminum blocks and in racing applications. *Image courtesy of Total Seal*

Similarly, a compromise must be made when determining ring thickness. A thicker ring will be stronger, less likely to flex, and more capable of dealing with heat and high pressure. But a thicker ring will also bring more friction with it, based on its increased sealing surface area against the cylinder wall. Ring resistance is one of the many areas being heavily researched now, as developments in ring design and material have allowed for thinner, low-resistance rings capable of dealing with the intense environment without flutter or twist. For street enthusiasts who are building their engines for long-term use and durability, the traditional thicker rings are a better bet. For racing engines that will be torn down and serviced regularly, the low-tension, thinner rings may offer a performance advantage for competition use.

There are many different ring profiles and materials available as well. The pros and cons of these different rings all boil down to what's best for your particular application. The piston rings serve to seal the piston to the cylinder wall to contain the pressure of combustion while containing the oil covering this same wall surface. A stack of three distinct and different rings has become the standard, with the top ring being most responsible for sealing, the second ring being responsible for both assisting the top ring with pressure control and assisting the bottom ring pack (called the oil rings) with oil control. The oil rings are a group of scrapers whose function is to clean the cylinder wall of excess lubricant.

Ring tension, profile, and even the spacing of the rings relative to each other are all subjects for discussion. I have always consulted both the piston and ring manufacturers before deciding on a final package for my engine projects, and I would have to recommend this to anyone investing heavily into a custom engine, regardless of its purpose. The

many varieties of piston ring styles on the market exist to fulfill different purposes, and no one knows more about the distinct advantages of each more than the piston and ring manufacturers themselves.

The bottom of the piston is typically defined by its skirts. These are the portions of the piston that extend downward from the deck and below the piston pin. The skirt serves to stabilize the piston within the cylinder bore, and it accomplishes this task through close tolerances. The larger and deeper the skirt, the more stable the piston will be in the bore. The downside to this is the increased weight and friction generated by the additional surface area of the skirt.

For performance enthusiasts, this offers a challenge. How small of a skirt can you get away with to limit weight and friction, while ensuring the skirt is large enough to adequately support the piston in the cylinder bore? Factory piston skirts have typically been generously large, as peak performance was not their intended function. Rather, quiet operation and long-term durability were the goals of the factory engine designers, and they added weight and friction to accomplish the goal. The most modern factory designs now boast smaller skirts, and some are even given a coating of a graphite to minimize friction and resistance. Pistons designed purely for racing at high rpm in short-stroke engines have virtually no skirt at all, and are amazingly light.

Running very short piston skirts will result in more engine noise through audible skirt slap. This phenomenon occurs when the piston rocks atop the connecting rod, causing an audible slap as it changes direction and the piston skirt comes in contact with the cylinder bore. This is more pronounced at low engine temperatures, before the pistons have had a chance to heat up and expand to full size, and also at low engine rpm levels, when there's less engine oil on the cylinder walls. Once the engine reaches normal operating temperature, most incidents of skirt slap disappear at normal operating rpm. If the engine's piston-to-wall clearances are properly set, skirt slap should not be an issue at normal operating temperature, even with the most minimal skirts. While it can be an annoyance, it's not typically a problem in high-rpm performance engines.

For street engines, the annoyance can be more pronounced with a quiet exhaust system, and I've known enthusiasts who have spent many hours trying to "fix" the noise. It can sound like a flaw in the valve train or even an exhaust leak, and if you're not used to hearing some skirt slap, it might fool you into thinking something else is wrong. In my experience, any engine designed for the street can sacrifice the minimal performance gain a short piston skirt would give. The addition of graphite coatings truly minimizes the friction loss, so go with more skirt and less noise. Having a quiet engine helps other problems (like valvetrain issues or exhaust leaks) be heard, found, and corrected before they cause major damage.

Chapter 4
The Camshaft and Valvetrain

The camshaft is generally regarded as the "brain" of the engine. Because so many factors change as a V-8 engine gains rpm, finding a single camshaft grind capable of performing well across a wide spectrum is truly a challenge of compromise. At low rpm levels, the air/fuel charge enters the engine at a relatively slow pace, and if it is able to travel through a small passageway and enter the cylinder through a small orifice (defined here as a valve that is not opened very far), the engine will operate efficiently.

Consequently, the same engine's need for more air and fuel at higher rpm levels requires that the passageway(s) are likewise enlarged. Larger intake ports, and valves that open farther for longer times, will enhance this same engine's efficiency at higher rpm levels. The need for a camshaft capable of operating efficiently over the wide rpm range has

prompted extensive research to determine the effect of each and every variable of the shape of a camshaft lobe. The size and shape of the lobe and its timed relationship to both its partnered lobe and the crankshaft all affect the engine's performance. Maximizing every possible variable in the design of the cam will serve to maximize its potential.

The best place to begin when choosing your camshaft is to consider the overall purpose of the engine you're building. A small-block V-8 powering a heavy car intended for cruising to the local burger joint will require a drastically different camshaft than a big-block V-8 engine designed to propel a lightweight drag car down the quarter-mile. Being honest with yourself is important here, because it's easy to choose a camshaft with too much lift and duration for your purpose. High-lift, long-duration camshaft and lifter kits cost about the same as low-lift, short-duration kits, and it's not hard for an enthusiast to dream big and choose the wrong cam.

Traditional American V-8 engines have, for the most part, been around for a while and have been heavily researched. It's inevitable that other enthusiasts have built engines very similar in design and purpose to your own, and a terrific camshaft is probably readily available from one of the many expert camshaft manufacturers in the aftermarket. Should your research prove this not to be the case, getting a camshaft custom-ground to your own specifications is neither impossible nor cost prohibitive.

The primary considerations when determining your own cam specs are the engine's displacement, compression ratio, and target rpm range. The cylinder head's breathing capability is also a primary consideration, and having flow-bench data for your heads before choosing a cam is a great advantage. If you know how big your engine is, how fast it will spin, how much cylinder pressure it will generate, and how much air the heads are capable of flowing, you're well on the way to finding the best possible camshaft for your application.

Knowing your target rpm range is essential throughout the entire engine design process, but this is especially true with the camshaft. The relationship between camshaft specs and the engine's power range is direct. For example, a cam with higher lift and longer duration will perform better at higher rpm ranges than a camshaft with smaller dimensions

Matching valvetrain components is a wise move, as the parts you choose will have been designed to work together as a system. Of course, many components from different manufacturers will work seamlessly together, but if you have any questions, getting parts from the same manufacturer will prevent fit or compatibility headaches.

Many would agree that the pinnacle of development for carbureted V-8s exists in NASCAR. This is an SB2-series engine from the Hendrick shop. Comfortable to over 9,000 rpm for extended periods and capable of making over 800 horsepower on high-octane gasoline, these 360-cube beasts are really something special. Unfortunately, they operate at such extremes that there's not much cam technology that can truly trickle down to street enthusiasts. The secret nature of these engines' camshaft development programs doesn't help either. Even if you did know the cam specs on this engine, you wouldn't want to use them in your street machine.

in an identical engine. The overlap between the intake and exhaust events offers improved performance at higher rpm levels, while simultaneously reducing efficiency at lower rpm levels. And again, this will be engineered into the camshaft, based on the target rpm window.

Because compromise is the key to designing the best possible engine to suit your needs, it's up to you to decide what you want the engine to do. If high-rpm operation is the goal, the camshaft and all the other parts going into the engine should be engineered for high-rpm use. If your powerplant is destined to spend more time on the street than the track, develop your engine to run efficiently at the rpm levels where the engine will spend the most time. While bigger camshaft dimensions translate into bigger power numbers in engines developed to use them, having the valves open too much or for too long will cost power in lesser mills.

This is where knowing the potential of the cylinder heads you've chosen really helps. Flow testing your cylinder heads will tell you how much lift they can handle, and where they flow best. By choosing your camshaft's dimensions based on the purpose of the engine and the potential of the cylinder heads, you'll be able to maximize the power of the design as a package. Other factors will assist you (and your cam grinder) in finding the best possible cam for you. The first consideration is the vehicle's weight. (A heavier

car needs to make good bottom-end torque to accelerate effectively.) Also consider the type of transmission you'll be putting behind the engine. (An automatic transmission needs to idle effectively when in gear, and loaded against the torque converter, while a manually shifted gearbox will be idling without any load as the clutch will be disengaged.) Naturally, there are some camshafts that would work acceptably in either application. (The cams GM uses in its immensely popular ZZ series of crate engines are a good example.) But power gains could be found if the owners feel the investment is worth it. If you're building an engine from the ground up, getting the best possible cam with the least possible compromise only makes sense, because you have to purchase a camshaft anyway.

For many years, flat-tappet lifters (both hydraulic and solid) were the only choice for V-8 engine enthusiasts. Since the development of the roller lifter (also in both hydraulic and solid form), things have changed dramatically. The roller's ability to follow a more radical camshaft profile has opened the door to performance levels far beyond the capability of traditional flat tappets. Amazingly, there has been no sacrifice in durability to achieve these gains, as even factory engines now use roller tappets (in hydraulic form) and carry long-term warranties with them. There is no reason not to run roller lifters anymore, unless they are simply not available for your engine of choice.

Roller lifters can offer better performance than comparable flat tappets' lobes because of their ability to open and close at a faster rate. The added expense of roller tappets used to be difficult to justify, but changes in oil additives have made premature wear with flat tappets an issue. Now more than ever, roller tappets are the best choice for any high-performance engine. Shown here is a flat tappet camshaft (left) next to a roller tappet camshaft (right).

LIFT AND DURATION

When choosing a camshaft, lift and duration are two of the primary variables. When looking for an optimal lift number, I ask the cylinder heads what they want. Flow-bench testing of your cylinder head of choice will tell you where the intake and exhaust ports are working best, and these numbers translate to lift numbers for your camshaft. If your intake port flows best at 0.500 inch of lift, for instance, and flow suffers past that point, why would you choose a cam with 0.600 inch lift (or more)? Research has shown that the more time the valve can spend at the optimal flow point, the more power the engine can make. Therefore, lifting the valve just past the optimal point while opening, and then letting it go past that optimal point again while closing will give the most possible time at that optimal flow dimension.

Stock GM Vortech Head

Intake Flow Chart

Valve Lift	0.100	0.200	0.300	0.400	0.500	0.600
Flow Meter	20	34	64	74	75	76
Total CFM	59	116	190	220	223	225

Stock GM Vortech Head

Exhaust Flow Chart

Valve Lift	0.100	0.200	0.300	0.400	0.500	0.600
Flow Meter	22	51	62	74	76.5	79
Total CFM	45	105	127	152	157	162

Ported LT1 Head

Intake Flow Chart

Valve Lift	0.100	0.200	0.300	0.400	0.500	0.600
Flow Meter	22	40	66	82	95	98
Total CFM	65	119	196	244	282	291

Ported LT1 Head

Exhaust Flow Chart

Valve Lift	0.100	0.200	0.300	0.400	0.500	0.600
Flow Meter	27	52	73	87	94	98
Total CFM	56	107	150	179	193	201

When discussing flow numbers, it's best to have a standard to compare your results to. Shown are flow test results for both a stock GM Vortech head and a ported GM LT1 cylinder head with larger diameter valves installed. The improvements can be seen in the increased flow numbers, but at what lifts those improvements occur may affect the camshaft selection as well. Because the Vortech intake port gains very little from 0.500 inch to 0.600 inch, there would be little benefit in choosing more than a 0.500 inch lift cam if this head is being used. Consequently, the ported LT1 head keeps gaining flow to the 0.600 inch lift point, so a cam with higher than 0.500 inch lift would be beneficial. Due to differences from flow-bench to flow bench, it's not advised to compare results from different flow benches. Flow test results are truly best compared to others done on the same bench. The "flow meter" numbers refer to the scale on the flow-bench fixture, and these are converted to cfm numbers based upon which flow scale is being used, and at what pressure.

To illustrate this, let's use the 0.500 inch figure once again. If you choose a cam with 0.510 inch lift, the cam will open the valve past the 0.500 inch optimal lift point to its maximum 0.510 inch number, then spend more time at 0.500 inch on the way down, thus optimizing the time when the port can do its best work.

The exhaust side works the same way; just make sure a pipe is attached to the exhaust port during flow-bench testing to simulate a header pipe. Simulating the actual route the exhaust flow will have to negotiate helps deliver a more accurate number, but there's no heat involved, so the data isn't perfectly accurate. One of the major forces working to get exhaust out of the engine is heat, as the expansion of the superheated, burned air/fuel mixture will be headed toward the path of least resistance. Also, the mechanical aspect of the exhaust flow (the gases are being pushed out of the cylinder by the piston) will skew the flow-bench numbers slightly, so you cannot take them as literally as you can on the intake side. I typically add 10 percent to the flow data on the exhaust side to account for these factors, and this number has served me well. I also add a dose of common sense to this, because the pipe used on the bench will probably not be of the same diameter as the header I end up using, and there will be resistance to exhaust flow in the rest of the system when compared to the flow bench.

Experienced high-performance engine builders like to see the exhaust side flow around 75–80 percent of what the intake side is capable of, and this is a good number in most cases. I certainly wouldn't like to see it any less than that, and because most of the engines I work with are destined for use on the street, having an exhaust port that flows closer to what the intake is moving is a good thing. Knowing the engine will have to exhale through a full exhaust system and mufflers means it'll need all the help it can get. The exhaust system design, from the headers to the tailpipes,

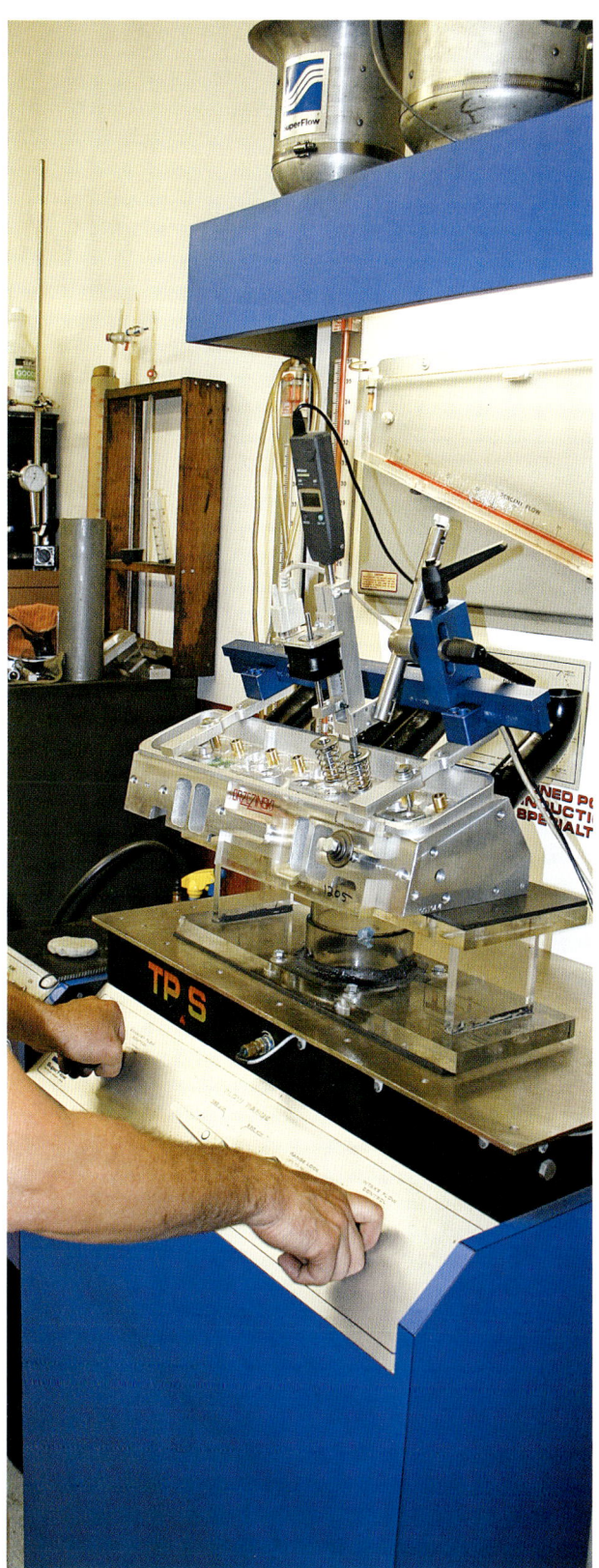

To choose the best possible camshaft for your engine, flow testing the cylinder heads is a must. Knowing the capabilities and limitations of the heads will help you choose a cam to work within these limits. It's difficult to choose a camshaft if you have no idea how your heads flow in the valve lift ranges you're looking at.

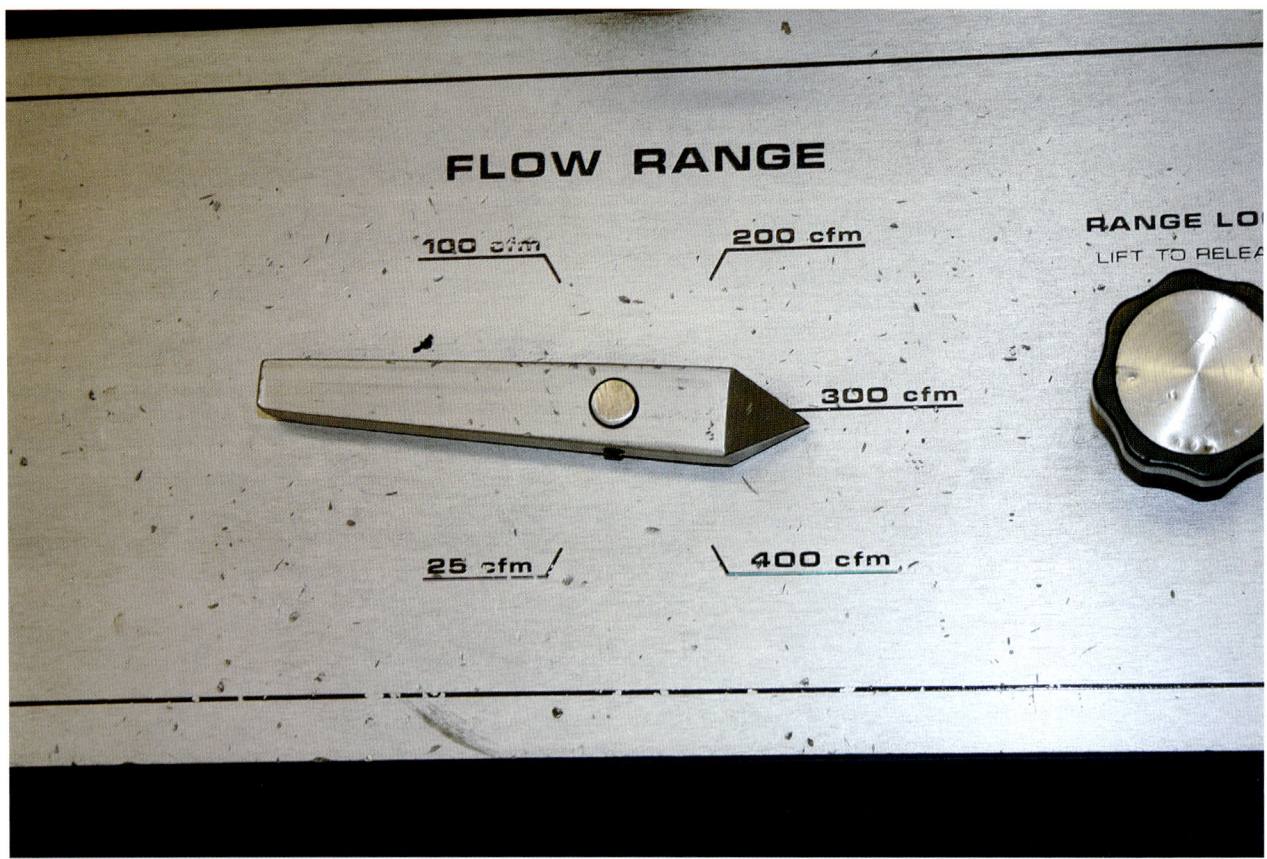

A flow-bench is also able to flow varying amounts of air to test flow quality at different lift points. Flow benches can also move air either into or out of the head to test both intake and exhaust ports.

can be executed to minimize back pressure and help power, but we're still talking about camshafts here. For camshaft selection purposes, have the exhaust side flow-bench tested, add 10 percent, and go from there when deciding on exhaust lift.

Naturally, without the airflow data from the flow-bench to guide you, you're just guessing what cam lift would be best, and that will probably cost you dearly in power production. Since you're trying to design the best possible engine, investing in flow-bench testing on your cylinder heads is a wise move. Once accomplished, you can select your cam lift with confidence.

Choosing the duration number for your cam is a bit trickier. In an optimal situation, the valve would open as quickly as possible, stay open at the optimal flow point as long as possible, and then slam shut as quickly as possible. The reality is that we want the valve to open and close smoothly, so the lifter can travel confidently up and down the cam lobe without excessive wear on any of the valvetrain components. What we end up with is a compromise, typically based on the chosen rpm range of your engine. The optimal duration grows as the target rpm range rises, and the longer the duration is, the more performance you'll sacrifice at lower rpm levels.

Camshaft duration is measured in degrees of crankshaft rotation, and is typically discussed in terms of duration at relative points of lift. The more duration there is at a typical lift point, the faster the valve is being opened (and closed). Short-duration cams make lots of low-speed torque and deliver crisp throttle response as a result. Longer-duration cams are best at higher rpm levels and make more top-end power.

Typical camshaft specifications give two measurement points: a dimension at 0.050 inch of lift and a total "advertised" duration figure. Short duration cams (typically with less than 220 degrees of duration at 0.050 inch of lift, and less than 300 degrees of advertised duration) are best for mild engines that require steady vacuum and smooth idle qualities. Once you get to 220–270 degrees of duration at 0.050 inch lift and approach 300 degrees of total duration, idle quality will get choppier and vacuum will be low, but the power band will head past 6,000 rpm. If you intend to push rpm levels past 6,500 and closer to 7,000, the cam duration at 0.050 inch lift will typically be in the 240-degree range or more with 300 or more degrees of advertised duration. Idle quality will be rough with hardly any usable vacuum signal.

Because V-8s are four-stroke engines, the crank turns twice as fast as the camshaft, so the cam gear is twice the size of the crank gear. To ensure proper timing between them, a pair of timing marks on the gears are aligned. These marks can be seen at the top of the lower crank gear and at the bottom of the upper cam gear. The alternative keyway cut into the crank gear is used when the builder wishes to alter the timing to either advance or retard the cam's actions relative to the position of the crankshaft.

If you've narrowed down your optimal lift and duration figures, it's time to consider the overlap and lobe separation specifications. Overlap describes what happens when both the intake and exhaust valves are open at the end of the exhaust stroke and at the beginning of the intake stroke. At low rpm levels, keeping the exhaust valve open a bit while the intake valve begins to open is usually a bad idea. The incoming air/fuel charge is contaminated with leftover exhaust gases, and there's a chance some of the intake charge could leak right out of the still-open exhaust valve before even being burned. However, once rpm levels increase, something very different happens.

As the exhaust valve is closing at high rpm, the gases are being drawn out at a high rate of speed. A well-designed set of headers really helps this scavenging effect, as the gases leaving other cylinders draw upon the cylinder with the open exhaust valve, increasing the efficiency of the exhaust stroke. If the intake valve is opened a bit early, this exhaust scavenging action will help draw in additional air and fuel on the intake stroke, having an effect almost like a supercharger. When done correctly, the engine's volumetric efficiency can approach or even surpass 100 percent efficiency, meaning that it exceeds the efficiency level that is determined purely by the displacement dimensions of the mechanical parts. Use of overlap in this fashion is only effective at higher rpm levels. As I mentioned, it will negatively impact engine performance at low rpm levels, so discretion is a must.

The amount of overlap a cam has is determined by how far apart the camshaft lobe centerlines are. These lobe separation angles (LSAs) are described in degrees, and the lower the lobe separation angle, the more overlap they will have.

A typical factory camshaft will have lobe separation angles of 112–114 degrees, with minimal overlap. High-performance cam grinds will typically have 110–112 degrees of lobe separation, while high-rpm racing cams range from 106 to 108 degrees of separation with generous overlap.

Some other considerations to make are more common sense. Naturally, a larger-displacement engine will require a larger (longer duration and/or higher lift) camshaft to feed it. A cam considered too big for a street-based 350 might be considered small for a big-block with another 100 cubic inches to feed. A mild cam for a small-block Chevy displacing 383 inches might be a radical cam for a 327-inch version of the same engine.

Once you've determined what you feel are the best possible camshaft specifications for your engine, it's time to ask the professional cam grinders for their opinions. Even after years of building engines of all makes and displacements for a wide range of purposes, I still call the premium cam grinders and get their recommendations and opinions. I get a great sense of accomplishment when there's a consensus about what cam dimensions would best suit my purposes, particularly if my cam ideas are aligned with the opinions of those who design camshafts professionally.

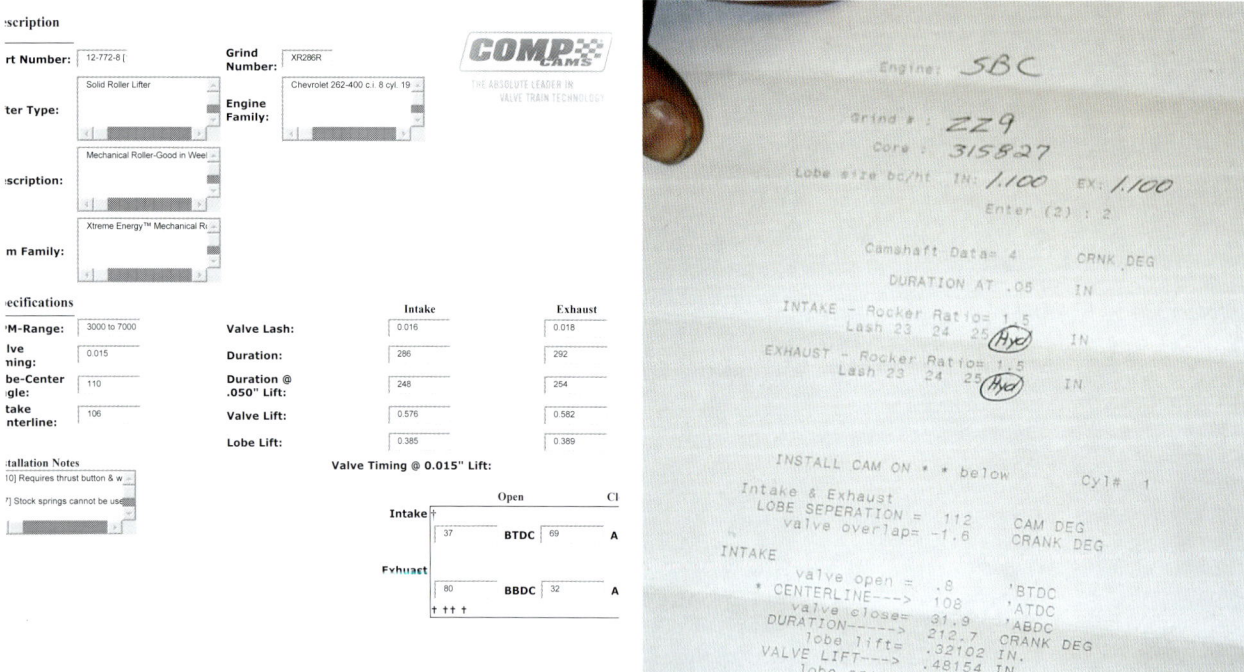

Here are two examples of typical cam cards that will ship with your camshaft. They will give you every relevant cam specification, along with general advice and important recommendations (like which valve spring to use with the cam). Never, ever throw out a cam card! You will find yourself referring to it often, although the information can easily be referenced online. The Comp Cams solid roller camshaft described on this particular form is one I run in my car. The printout in the photo is the cam spec sheet for the popular ZZ9 SBC camshaft profile from TPI Specialties.

Also, chatting with cam experts about their thoughts and bouncing ideas off them only increases my own knowledge and level of understanding.

If I'm building a marine engine, I speak with marine engine experts. If I'm building a street engine, I speak with street engine experts about their experiences. There is absolutely no shame in seeking out the opinions of others, especially if they have more experience. There have been many times when I thought I'd made my final decision, and my mind was changed after speaking to someone who had more experience. I usually narrow my choices down to two cams I know will work really well, and finalize my choice after speaking with experts. I wholeheartedly recommend you do the same.

The camshaft in a traditional V-8 engine is designed to run at half the speed of the crankshaft, and the synchronicity between the two results in camshaft timing.

The camshaft's actions relative to the piston's position is critical to power production. Camshaft designers carefully engineer the valve openings and closings to happen at specific times, depending on where the piston is in the cylinder. The most critical of these valve motions involves the closing of the intake valve. As the piston moves downward, the intake valve should rush to open and allow the incoming air/fuel mixture to be drawn into the cylinder. Once the piston has reached BDC, the valve must close completely before the piston begins moving up into the compression stroke. Should the valve close too early, the full potential of the intake stroke will not be realized. Should the valve close too late, the air/fuel charge will begin to be pushed out of the intake valve again as the piston begins moving up the cylinder.

Engine builders and tuners have the capability to alter the timing between the crankshaft and the camshaft, but this ability comes with consequences. Altering the cam timing so valve events happen sooner (advancing the cam) means the power and torque curves will also occur sooner, at lower rpm levels than when the cam is installed as designed. Similarly, delaying the valve's opening points (retarding the cam) means the engine's power peaks will occur later, higher in the rpm range. Why would an engine builder want to do these things? Based on the track the engine will be performing on, cam timing can be used to move the torque and horsepower peaks to a more advantageous place. For instance, if a dragster is having traction problems, the cam can be retarded a bit to kill some low-end torque and move the power band higher in the rpm range, where the car has sufficient traction. In this same vein, if a stock car has plenty of traction coming off the corners and could use more power there, advancing the cam slightly will bring the power band down a few rpm and deliver more torque to the tires as the car comes out of the corner.

The cam timing can be moved if the engine is equipped with an adjustable cam gear like this one. The slotted holes allow for quick and easy cam timing adjustments, and the graduated markings on the gear make it easy to see how much you're moving things. Remember, changing the cam timing will change the relationship between the valves and the pistons. If your piston-to-valve clearance is tight, make sure altering the cam timing won't cause any problems. This is one of the many things to check while building the engine. By checking the piston-to-valve clearance at varying cam timing points, you'll know how far you can go with this adjustment.

It should come as no surprise that moving the cam timing away from its optimal position will result in a loss of peak power too. But in some scenarios, engine builders and race car tuners gladly sacrifice some power to move the peaks to where they can benefit the car more. Power is useless if you can't use it, and savvy tuners can make their cars faster by killing some power to gain traction.

I mentioned there could be consequences when altering cam timing, and it's not hard to see how problems could arise. The piston-to-valve clearances in high-performance racing engines are always close, and moving the timing between the valves operating and the piston moving can result in piston-to-valve interference. It's an easy way to bend (or break) intake or exhaust valves, damage a piston, and ultimately destroy an engine. This is especially true in high-rpm engines, where parts are moving so quickly they actually stretch. Connecting rods are well known for this, and the combination of a stretching rod and altered cam timing can cause parts to make contact even if they had clearance when the engine was being turned over by hand. For street-driven applications, there should be no justifiable reason to alter cam timing away from manufacturer's recommendations.

To set the initial camshaft timing, a degree wheel is attached to the crankshaft. The engine is turned over slowly by hand, and the cam lobes are checked to ensure each of the valve actions occurs at the proper time relative to the crankshaft (and therefore the piston). Clearances between the valves and pistons can also be checked, but the primary reason for degreeing a cam is to ensure the valves are opening and closing at precisely the right time. Naturally, I base my cam degreeing efforts first on the critical intake valve closing point, and once this is set, I then double-check the rest of the valve actions.

All of the camshaft's critical dimensions should be referenced on the cam card that ships with every new camshaft. Each of these dimensions can be checked with a degree wheel in place, and should any (or all) of the dimensions be off, the cam-to-crankshaft relationship can be altered to bring the relationship between them back into sync. There are various methods to accomplish this, from offset cam dowel pins to slotted camshaft gears and more. For street enthusiasts, the odds are good you'll never have to alter cam timing at all, as today's cam manufacturers offer precision grinds and terrific quality control to ensure the timing of each cam event is right on. If your degreeing exercise finds this not to be the case, exchanging your faulty cam for a new one should not be a problem.

Using the degree wheel may be a bit intimidating at first, but once you set the degree wheel in position with TDC of the number one cylinder, it starts to get easier. A 1-inch dial indicator is placed on the intake lobe of the same cylinder, and by rotating the crankshaft (and degree wheel) to the

Once the camshaft is installed, it's essential to check the relationship between the crankshaft and camshaft for proper timing. To accomplish this, a degree wheel is used. It is attached to the crankshaft and a pointer is used to compare the camshaft lobe actions to the degrees of rotation of the crankshaft. To begin, the number one piston is brought to TDC. The timing gear marks should be aligned, and this bridge tool is commonly used to ensure the piston is at the very height of its travel.

Once TDC has been verified, the pointer should be adjusted to point to zero on the degree wheel.

The first dimension commonly checked is the lobe center. This will establish that the number one cylinder's intake valve will be open at the correct time in relation to the position of the number one piston. Rotate the crankshaft until the lobe opens fully, and zero the micrometer at that point.

Rotate the engine until the micrometer moves 0.020 inch, and mark the degree wheel. Then, rotate it back in the other direction, back over the nose of the cam lobe, and continue until the micrometer moves 0.020 inch down again. Mark the degree wheel once again.

Subtract the two numbers and split the difference to find the number exactly halfway between them. This represents the absolute peak of the lobe, and should coincide with the cam card dimension for lobe center. In this case, the lobe center dimension is 108 degrees, so we're right on. If these numbers correspond, you know the cam is installed in the proper relationship to the crankshaft in this regard. It is possible to check every timing dimension ground into the camshaft using the degree wheel and micrometers in similar fashion.

A wide range of lifters are available in the aftermarket, each with its own characteristics. The odds are good you won't require a specialized lifter unless you're doing something really strange. Even then, you might be able to use an existing lifter designed for something else. The roller lifters in this photo represent Crower's offerings, and as you can see, they cover small- and big-block Chevrolets, LS-series GM engines, Fords, Chryslers, Buicks, and Pontiacs.

In addition to the rounded lobes on the roller camshaft (right), note the machined step in the nose of the cam. Roller cams must be secured in the block to prevent them from moving forward and backward while the engine is running. The step in the front of the cam is how GM chose to retain them, and it corresponds with a lock plate used in later-model factory blocks that came with roller cams.

prespecified points referenced on the cam card, and checking the measurements on the dial indicator, you will soon make sense of the various reference points.

Flat Tappets vs. Roller Tappets

If you've never run a roller lifter before, you're in for a sweet surprise. The capabilities of the roller lifter design far outweigh those of the flat-tappet lifter, and the only reasons to run a flat-tappet lifter anymore are budgetary constraints, rule limitations, or to maintain a level of originality in the engine. Roller tappets can open and close faster and do so with less internal engine friction, so the lobes they roll across can be more aggressive without sacrificing anything else. The recent concerns over zinc and ZDDP levels in modern oils are limited to flat tappets, as roller tappets do not maintain metal-to-metal contact like flat tappets. I can confidently say I will never again choose to build an engine with flat tappet lifters, unless a client insists upon it.

With that being said, the only question left is whether to run hydraulic roller lifters or solid roller lifters, and this is typically an rpm-related question. While new parts assure more, the overwhelming majority of hydraulic roller lifters cannot perform reliably past 6,500 rpm. Hydraulic lifters are designed with an internal oil chamber and a floating pushrod socket, and they use pressurized engine oil to provide a hydraulic cushion between the pushrod cup and the body of the lifter. This allows the valve lash to be zero when properly adjusted. However, the rigors of high rpm may begin to collapse the hydraulic lifter, resulting in less than full opening of the valves. In most V-8s, if you intend to run more than 6,500 rpm, a solid roller lifter is your best choice. If you prefer minimal maintenance (valve clearance adjustments) and don't plan to spend much time past 6,500 rpm, the hydraulic roller lifter is a fine choice. Personally, I don't mind adjusting valve lash occasionally, and I want every bit of power available to me, so I really like solid roller lifters. I'm never left wondering if more power can be had, and I'm not scared to push the engine to its rpm redline at will. That's what high-performance engines are all about.

Roller lifters do require adequate oiling, however, and the latest generation of roller lifters has been engineered with increased oiling capabilities. There used to be concerns whether true street engines, which operate at idle and low rpm most of the time, could sufficiently lubricate roller lifters. This is no longer a concern, though, and I've not seen a roller lifter fail because of insufficient lubrication in several years. The only remaining concern about roller lifters is their need to be rebuilt occasionally, and if they are not, they may wear out and break. However, the premium-quality roller lifters being offered today shouldn't require inspection and/or rebuild for

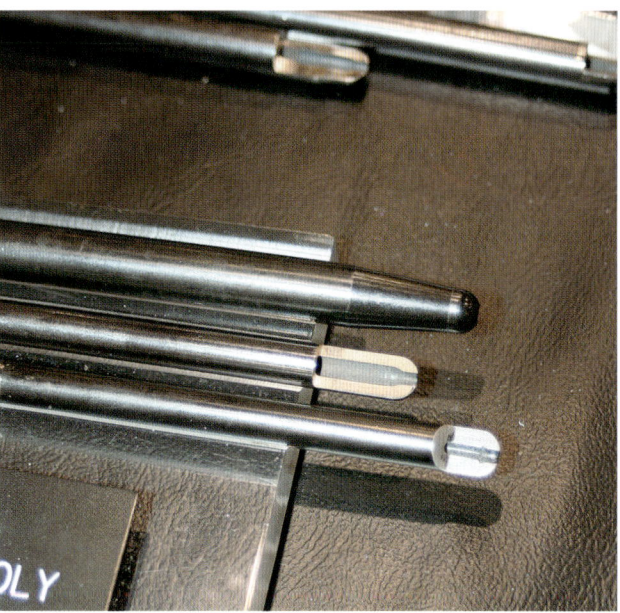

Here's a look at some top-quality pushrods. While their shapes may vary, all have over 0.100 inch wall thickness. These are Manley products, and have proven themselves in many forms of racing. Pushrods are available from many sources, and assuring that they are made to withstand the rigors of the track is a wise move.

at least 20,000 street miles, so it's nothing to worry about until your engine has served you for quite a while. Still, it is a concern, and if you have any questions about the wear on your lifters (if you're purchasing a used engine equipped with them, for example), it would make good sense to return them to the manufacturer and have them rebuilt. This procedure is less expensive than buying new tappets and is cheap insurance. Knowing your roller lifters are fresh and ready for duty will add to the confidence you have when flooring the gas pedal, and I'm a big fan of confidence.

Pushrods

The pushrods are deceptively simple. Most see them as basic components and of minimal importance in a high-performance engine. This is simply not the case, as the pushrods transfer all of the lift and duration from the camshaft to the rocker arm, and any amount of bend or deflection will directly result in lower performance from the engine.

I like to run a beefy pushrod to minimize deflection and keep the potential to bend at bay. A performance engine will inevitably require a strong valve spring, and the greater the tension on the valve spring, the greater the pressure on the pushrod.

Designwise, the pushrod can be described as a loaded column, as it is basically a tube that must support a load at the top and at the bottom. (This is true at least in typical designs where it is hollow, and oil flows through it like a straw.) The pushrod is not always loaded in a perfectly straight line from top to bottom, however. There is typically some side load on any pushrod. If the pushrod is angled at all, like in an offset rocker or lifter application, these forces can be great.

I like to use the largest-diameter pushrod that will comfortably fit into the engine, which means the pushrods have at least 0.010 inch clearance to any other component at any point in their travel. This dimension becomes critical when pushrod guide plates are being used, since having too much clearance may allow the pushrod to deflect too much. Guide plates are designed to keep the pushrod correctly aligned, and having too much room around them will obviously prevent them from doing their job.

In engines without guide plates, other engine parts (like the cylinder head casting itself) can be used like a guide plate if the 0.010 inch dimension is maintained.

Many engine builders feel that running a thicker-walled or larger-diameter pushrod will add undue weight to the valvetrain. I feel that because the pushrod is on the "light" side of the rocker arm (rather than the spring-loaded, longer side of the rocker), and because beefy pushrods weigh only slightly more than the stock parts they typically replace, they are well worth the investment and security.

Pushrods should be checked for straightness before installation, and this simply involves rolling them on a known-flat surface (like a piece of glass). I roll two or three of them together at once, and if one is not perfectly straight, you'll feel it. Once they are straight and have been cleaned out thoroughly (ensure hollow push rods are free of any debris internally), they should be ready for installation.

It's critically important to have the proper-length pushrod. The pushrod locates the rocker arm tip on the top of the valve stem. When it is correct, the tip of the rocker will travel across the top of the valve without getting near the edge in any direction. This can be checked by marking the top of the valve

The pushrod guide plate keeps the pushrod aligned with the rocker arm. There should be adequate clearance all the way around the pushrod when the engine isn't running, but some evidence of minor contact (as seen on the pushrod) is normal.

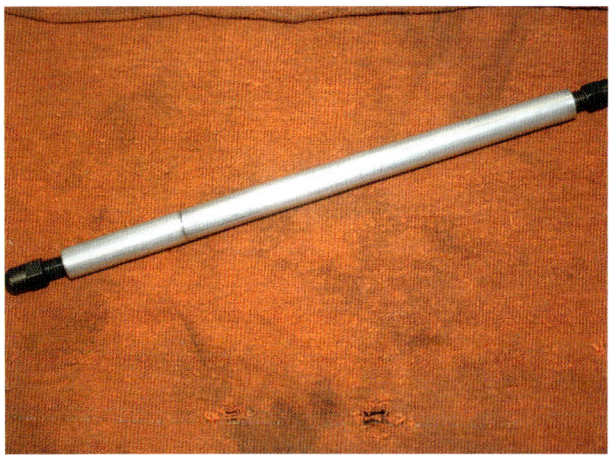

An adjustable pushrod allows the builder to determine the proper pushrod length to ensure proper valvetrain geometry. Both ends are threaded, so they can be easily adjusted.

This dummy rocker arm from Manley makes finding the pushrod length easy. You can see the adjustable pushrod has been installed between the lifter and the dummy rocker arm. The pushrod is extended until it makes contact with the flat surface of the tool, which is also making flush contact with the valve tip atop the spring.

To ensure proper rocker arm contact, the proper-length pushrod is installed along with some soft springs, and the valve tip is colored with a marker.

The rocker arm is installed and adjusted, and the engine is rotated to move the rocker arm through its entire range of motion.

The rocker arm is then removed, and the pattern left in the marker will show how well the rocker arm is aligned with the valve tip. The pattern should be centered across the valve tip from top to bottom and side to side, as shown.

Aftermarket manufacturers typically stamp their products with the rocker ratio for quick reference. A glance at these rocker arms lets us know they are a 1.60:1 ratio.

Three basic rocker arm types are commonly found on domestic overhead-valve V-8s. Stock-type rocker arms are either mounted on a stud or a shaft, with studs being the more common of the two. Factory stamped-steel rocker arms are neither precise nor capable of handling the increased spring that pressures performance camshafts demand, so stronger steel or aluminum rocker arms are offered as upgrades for stud-mounted designs. For the ultimate in stability, stud-mounted rockers can, in many cases, be upgraded to custom shaft mounts with full-length bearings to ensure durability and precise control. Shown are stock-type rockers (from a GM LS-series engine), aluminum stud-mounted arms with roller tips, and bolt-on shaft rockers, also with roller tips. These aftermarket upgrades are all from Crane Cams.

with a marker, installing the rocker, and turning the engine over so the rocker travels through one complete opening and closing motion. Then the rocker can be removed again and the marker should be cleaned off where the rocker contacted it. The clean area should be centered on the valve tip, and should be no closer to any one edge than any other. This proves the rocker tip is traveling completely across the top of the valve, and is properly aligned. If the pushrod is too short or too long, the rocker tip will not be traveling over the valve correctly, and this will be apparent in the mark. In some cases, valve lash caps can be used to correct minor deficiencies, but getting the pushrod length correct the first time is the goal here.

I like to use a single, adjustable-length pushrod to determine the optimal dimension. I make sure to check both intake and exhaust valves on both sides of the engine to ensure everything is equal. Improper block or head surface machining may result in a difference from side to side, and improper cylinder head machining in the valve seats or on the valves themselves can result in a difference between intake and exhaust valves. Checking both an intake and an exhaust valve on both the left and right sides of the engine ensures your pushrod dimension is consistent.

Rocker Design, Ratios, Geometry, and Materials

The rocker's job is to transmit the motion of the camshaft to the valve. Like a teeter-totter, the rocker simply moves back and forth over a pivot point. The rocker's ratio is determined by the measurement from the center of the pivot point to the center of the pushrod socket compared to the dimension from the center of the pivot point to the center of the friction surface traveling across the top of the valve stem. These two dimensions will be different, and they are expressed as a ratio, with the distance on the pushrod side being the standard, and the distance on the valve side (which will always be larger) as the comparison. For example, a 1.5:1 rocker ratio means the distance from the pivot point to the rocker tip on the valve side will be 1.5 times the distance from the pivot point to the center of the pushrod socket. These ratios are typically referred to by the ratio number alone. If someone refers to "1.65 rockers," they are talking about 1.65:1 ratio rocker arms.

The reason rockers have ratios at all is to modify the lobe dimensions from the camshaft. The lift at the camshaft's lobe is transferred through the pushrod to the rocker arm, and then is multiplied by the rocker arm's ratio before being transferred to the valve. For example, a lobe with 0.400 inch lift and a 1.5:1 ratio rocker arm would open its valve 0.600 inch (0.400 x 1.5 = 0.600) The same lobe with a 1.6:1 rocker would open the valve 0.640 inch (0.400 x 1.6 = 0.640), and with a 1.65:1 ratio rocker, the valve lift jumps to 0.660 inch (0.400 x 1.65 = 0.660). For this reason, it's important to determine what rocker arm ratio you'll be running while you're deciding on a camshaft. All of the camshaft dimensions you see are based on a given rocker ratio, and should you choose a different ratio than the one used to develop the specifications published by the cam manufacturer, your final specifications will also be different.

Among the latest developments in rocker technology are these bolt-on shaft rockers from Crane. Designed for small- and big-block Chevy engines (with others to come, we hope) these use a plate bracket that bolts into the factory rocker stud holes in the cylinder head. Then the rockers and shaft are bolted to the bracket. The wide surface area of the shaft setup offers stable operation to high-rpm levels, and rocker deflection is minimized.

If stud-mounted rockers are to be kept on an engine running an aggressive cam or being pushed to high rpm levels, a good quality rocker arm is important. Shown are Crane's Gold rocker arms, which have proven themselves as a top-notch upgrade for decades. Naturally, they are available for many makes and models in many rocker ratios.

If budget is an issue, options for good rocker upgrades exist for most engine families. This Comp upgrade stud-mount rocker still uses the factory method of adjustment (with the rounded insert and a simple nut on the rocker stud), but offers a roller tip, better quality control, and more strength than the factory part it replaces.

This particular Comp rocker is designed to be self-aligning, with discs on either side of the roller tip keeping it centered. This means no pushrod guide plates will be required. Some newer heads don't have provisions for guide plates (like the popular GM/Chevy Vortec V-8s), so the self-aligning feature is a necessity.

When stud-mounted rockers are retained in a racing application, a stud girdle is a popular upgrade. Designed to support all of the studs and prevent flex, stud girdles have proven their ability to increase the stability of stud-mounted rockers and prevent breakage in extreme situations. Shown is a big-block Ford setup from Trick Flow racing for use on its aluminum 429/460 heads.

For the ultimate in valvetrain stability, a full aftermarket shaft rocker upgrade is the way to go. Shown is a T&D setup on a Pontiac V-8. Note the beefy base stand between the rockers, with the shafts bolted to them. You might also spot that these rockers are offset on the intake side. These clear much-widened intake ports, and are teamed with offset lifters to get around the widened ports.

So, what rocker ratio is best for you? For street use, I prefer to run a gentle rocker ratio. The greater the rocker ratio, the more dramatic the impact on the rest of the valvetrain components will be. With increased rocker ratio comes more stress on pushrods, rocker mounting hardware, valve springs, and even valve seats. A more aggressive rocker ratio will also be capable of increasing effective valve lift and duration, and will increase the rate at which the valve is opened and closed. These changes can have a positive impact on power production, but is the added stress worth the power gain? For a competition vehicle, I would say absolutely. If you're racing, you want every bit of power you can find, and if a higher-ratio rocker arm can deliver it, it's worth it. A competition engine will typically be checked more often for wear and tear, and the stresses that added rocker ratio brings with it can be found before they cause damage. The valve springs will be subject to additional stress, and will wear out faster than they would with lesser rocker ratios.

If street performance is your goal, and you might only pull the valve covers once or twice a year to check valve lash, then running a more conservative rocker ratio is the best advice I can offer. In a street engine, reliability and durability are much more of a priority than in a racing engine. Racing engines are expected to last a limited amount of time (sometimes only the duration of the race) and street engines are expected to last much longer. Surely, well-built race engines in some competition classes may last an entire season without requiring any repairs, but street engines are expected to last many years without major hassles. For street engines, I typically use the factory's rocker ratio choice. Additionally, I always consult with the camshaft and rocker arm manufacturer(s) and get their input on my choices before purchasing anything.

The rocker system is purely mechanical, and strong forces are at work here. The camshaft forces the lifter upward, and its motion is transmitted through the pushrod to the rocker arm. The rocker pivots, transmitting this energy to the top of the valve, which is being held closed by the pressure of the valve spring. The amount of pressure in the entire valvetrain system is based on the valve spring, and these pressures are transmitted to each and every component. The valve spring's pressure is enhanced by the rate at which the valve is being opened. This rate is controlled by both the camshaft ramp angle, the rocker ratio, and of course the engine's rpm.

Many factory rocker designs are based on a stud-mounted rocker arm. This simple design has proven effective in many performance applications, but there's never been any question that a given amount of flex occurs here. The amount of deflection, and its impact on performance, is not clearly defined by precise measurements. The alternative is shaft-mounted rockers. This design is hardly new (many factory engines use shaft-mounted rockers, including Buick V-8s, Mopar B and RB big-blocks, and Mopar Hemis). Shaft mounts do offer more support and stability than studs. The problem with shaft-mount rockers is that all of the rockers on a shared shaft will be located at the same height. If this height is within the proper adjustment for correct rocker geometry, there's no problem. But if the shaft is at the wrong height for one (or more) of the rockers mounted upon it, performance will be compromised.

Adjustability can correct all of these issues, however. While the bulk of the factory shaft setups offered no adjustment, they also were typically offered with hydraulic lifters, which have a relatively wide range of slop in them. The automotive aftermarket quickly developed adjustable rockers for the shaft setups as solid lifters became popular. The stud-mounted rockers have always been adjustable individually, so this was never an issue in the engines equipped with them. The aftermarket responded to the needs of stud-mounted rocker owners with much stronger rocker arms and much improved adjusters to hold the preferred clearance settings.

To enhance the durability of stud-mounted rocker designs, stud girdles were engineered to tie the studs (or more accurately, the rocker arm adjuster nuts) together once all adjustments were finalized. These stud girdles assisted greatly in adding longevity and durability, but naturally they added expense and made routine rocker arm maintenance more difficult. Additional valve cover clearance is also required when stud girdles are used.

Aftermarket manufacturers developed shaft-mounted rocker upgrades. The most commonly known shaft rocker upgrades were developed by Jesel and T&D Performance. While horsepower claims were made, enhanced long-term and high-rpm durability gains truly justified these upgrades. Stud girdles were no longer required, and the additional stability provided by the surface area offered by a wide shaft versus a stud cannot be argued. Deflection is a nonissue with shaft mounts—especially the well-engineered aftermarket units designed for racing use. These units typically use the former rocker stud mount holes to install, so no cylinder head modifications are necessary to install them. They are pricey, but the level of security and durability they offer can justify the expense. When you consider the cost of top-quality studs, aftermarket rocker arms, and a stud girdle, the price of upgrading to shafts isn't too much of a stretch. This is especially true for those running endurance racing engines. While shaft-mounted rockers might be unnecessary for a street engine, any racer pushing more than 7,000 rpm for extended periods should certainly consider them.

If stock-replacement-type rocker arms are your choice, you've still got some options. The tried-and-true aftermarket stock-replacement arms typically have roller tips, which eliminate the metal-to-metal friction found in most stamped factory designs. The reduction of friction in this critical area makes for a smoother rocker action and also reduces the heat created by the friction. As a rule, the factory-stamped arms are notoriously inconsistent in their ratios. The aftermarket

Shown is a matched valve spring, retainer, and a pair of keepers. This particular spring is a high-performance, triple-coil design, with two true springs and a dampener. Such a spring package would be capable of high-rpm operation for extended periods, even with a high-lift, long-duration cam with an aggressive lobe profile. The stainless-steel retainer and hardened steel keepers are capable of working in the harsh high-rpm environment as well.

Farther research into valve spring design has resulted in the development of so-called beehive springs (left). With their changing coil diameter, the harmonics are less of an issue and the beehive spring is capable of working effectively into the higher rpm ranges with less spring pressure than comparable traditional valve springs, like the one on the right. Naturally, matching retainers are required for the beehive springs.

arms are typically made of extruded aluminum, which by its nature delivers a more consistent ratio. All of the arms are made from the same extrusion, and this method is preferred over the relatively inexpensive stamping process.

Another option available when shopping for improved rocker arms is a roller trunion. This is a series of roller bearings engineered into the pivot point of the rocker, which eliminates another metal-on-metal friction point. Again, the heat generated at the pivot point is reduced. While these bearings can wear over time, they can also be rebuilt or replaced. I've never had one of my own rocker arms wear out, but I've seen competition engines where it's happened. There's normally plenty of warning when a rocker arm trunion bearing is wearing out, as it will make noise. If these warnings go unheeded, the bearing could fail, but as I mentioned, these occurrences are rare indeed. For a street performance application, the addition of good-quality roller rocker arms may not make more power (unless they carry an increased rocker ratio, as discussed), but they will surely generate less heat and offer more accuracy than the parts they replace. The extruded aluminum arms are also stronger, resulting in less deflection under the stress of a stiff valve spring. This alone can justify the purchase if you're planning an aggressive camshaft with a matching stiff spring. This is especially true when upgrading to roller lifters, as roller camshafts have more aggressive ramps than comparable flat-tappet versions.

Valve Springs

Valve springs are another integral part of the valvetrain. Some would even say the valve springs are the most critical component in the valvetrain. They are tasked with a very difficult job (working against the mechanical push of the cam to keep the valve closed) and they must do this work repeatedly in a harsh environment. If the spring is weak, the valve will not close completely, causing a loss of performance (at the very least) and potential engine damage if the still-open valve contacts the moving piston. The results of valve float are never pretty, and getting the right spring in place is critical.

So, why don't engine builders simply put the biggest, strongest springs in place? Remember, it takes energy to push against the spring and open the valve. The more energy this takes, the less power the engine will make. Choosing the right spring used to be a matter of finding the weakest-possible spring still capable of keeping the valves from floating. The problem with this theory is that springs lose some of their tension over time. While they may work just fine when new, they could lose tension over time and become weak enough to let the valves float eventually. Luckily, research has provided a

This stamped-steel timing cover from Comp Cams might not look like anything special, until you turn it over. A reinforcement has been added right where the front of the cam will line up. Because roller cams have a tendency to "walk" forward in the block, they must be restrained. The reinforced timing cover works in concert with a thrust button on the front of the cam to limit the camshaft's ability to travel forward in the block. Without this reinforcement, the cam will drive forward and bend the timing cover, which can cause internal engine damage. If you upgrade to a roller cam, get a cam button and a reinforced timing cover to match!

This competition-quality timing belt offers easy access and adjustability through slotted holes in the camshaft gear. The cog belt is ultrastrong, and absorbs any potentially damaging harmonics, preventing them from transferring from the crankshaft to the camshaft.

remedy. Additionally, springs work like tuning forks, in that they have a resonant frequency at which they lose tension. When they reach the point in engine rpm where this natural frequency occurs, tension is lost and valve float can occur.

A typical valve spring is a wound wire coil. To combat resonant frequencies, engineers added a second spring, wound in the opposite direction, inside of the original spring. The fit between the internal and external springs ensures they touch, in what is called an interference fit, which combats the resonance. Naturally, two springs have more tension than one, and dual springs have been the standard in high-performance V-8s for the last couple of decades. The downsides were that the springs rubbing against each other produced a large amount of heat, which served to weaken them over time. It's common for valve springs to break after hard use, and a broken valve spring cannot work as it's supposed to.

Research at GM produced a new alternative. Engineers found that if the diameter of the spring changed from bottom to top, the resonant frequencies could be eliminated without the need for a heavy second spring. Called "beehive" springs for their distinctive tapered shape, they offer improved performance and higher rpm capability with lower spring loads than traditional parallel-wound springs. Now they have become common in the aftermarket, and are recommended highly in many performance applications.

High-performance valve springs are rated and recommended on the amount pressure they exert, both with the valve open (called seat pressure) and the pressure they

exert when the valve is completely open. The recommended height of the spring when installed (called the installed height) is what the seat pressure is based on. It's important to ensure the spring is at the correct height once installed to guarantee the pressure is within recommended limits.

If a roller cam and lifter setup is selected, it is a requirement that the cam be retained in the block to prevent it from moving forward while the engine is running. This can be accomplished with a retaining plate on the front of the block, if the front of the cam is engineered to accept it, or by using a cam button mounted in the center of the cam gear. Such cam buttons are engineered to touch the timing cover once it is installed. The timing cover must be reinforced, typically through use of a doubler plate welded inside of it, to accommodate the friction of the cam button.

Beyond the timing chains shown, timing belts and timing gears have also been engineered to join the crankshaft to the camshaft. Timing belt setups offer reliable quiet operation and typically offer slotted holes on the cam gear for quick and easy cam timing adjustments, but this can be an expensive upgrade. Timing gears, while not terribly expensive, do offer precise cam timing. They are known to be noisy however, and any harmonics or vibrations in the crankshaft can transfer to the camshaft more readily than through a comparable chain or belt. For most street applications, the use of a top-quality timing chain set is perfectly adequate. In racing applications, many engine builders choose belt drives, due to their simplicity and ease of adjustment.

Chapter 5
The Cylinder Head

The V-8 cylinder head is arguably the most important part of a high-performance engine. The capabilities and characteristics of the cylinder head have a direct impact on all facets of power production. Luckily, we live in a time when improved cylinder heads are (or will soon be) available for all the popular American V-8 engines. While the best factory cylinder heads are still a great choice, especially when properly ported and machined by an expert, these heads typically came only on the best factory cars, and as such are in demand from restorers and collectors. This drives the price of these castings beyond the typical budget of the hot rodder or hobby racer. By purchasing aftermarket cylinder heads, the enthusiast gets a head start (because aftermarket head designs are engineered for improved performance and durability) and a weight break (because most aftermarket heads are cast in aluminum). The factory-design shortcomings are typically addressed, and performance features like improved port designs, larger valves, and more efficient combustion chambers are part of the deal.

Because we are discussing all American V-8s, I will try to keep this information general. Each engine has its own quirks and personality, and most of those differences are caused by the cylinder head design. You cannot build a Pontiac 455 in the same way you build a Chevy 454 or Ford 460, but some basics are universal.

First, the head must be capable of feeding the displacement of your chosen V-8. Factory engines were designed to work well at low rpm levels, where large intake and exhaust ports aren't optimal. It's rare to find a factory cylinder head where the ports are too big as cast. Most of the time, the ports will benefit from being enlarged to allow a greater quantity of airflow into and out of the engine. But maximizing engine performance isn't just about the quantity of air moving in and out of the engine, but also the quality of that airflow. An efficient port will move a good volume of air through it at high velocity, and years of experience have shown hot rodders where the trouble spots are.

This cutaway of a small-block Chevy head shows the basic design of an intake port. Note how the relatively large port opening, at left, tapers down and changes shape as it flows toward the valve seat. This is the challenge of developing a good intake port, as the flow characteristics change with the shape of the port.

This cutaway head shows yet another challenge—getting the incoming air/fuel past the valve guide. The valve guide and stem take up a lot of valuable space inside the intake port, so minimizing the size of the guide and grinding it to a more aerodynamic shape add to the flow capability of a head.

Vintage American V-8s don't always offer awesome performance as they were manufactured. This early Olds V-8 head is a great example. The combustion chambers are deep and shrouded, the valve diameters are relatively small, and the center two exhaust ports are siamesed, which hurts performance. This head can be made better by adding some modern influence to the chamber design, adding new high-flow valves in larger diameters, and extending the split between the center exhaust ports out to the exhaust flange to isolate them from each other.

As both the intake and exhaust ports are made more efficient, they typically become larger, because material is removed. A port that is too large will be inefficient, especially at lower rpm levels. By focusing on the efficiency of the port, the optimal result will be the smallest port size capable of feeding the engine adequately at its peak rpm point. Research has shown that while a larger port may make more peak power at the highest possible rpm level, a smaller, more efficient port will make more power throughout the rpm range, resulting in a better-performing vehicle overall. The key is in developing the best possible, most efficient port for the displacement of the engine being built.

The cylinder head design must work in concert with other components and factors, such as the valve size and the camshaft, to maximize the effectiveness and efficiency of the overall engine. Of course, the intake manifold is also part of the overall equation, but I typically work on developing the cylinder head and camshaft combination before addressing the intake manifold.

When choosing a cylinder head, its overall capability must be taken into consideration. This is where research really pays off, as the results of others who have built similar engines should influence your decision. Unless you're developing a combination no one has ever considered before, you can learn from others. For instance, some factory cylinder heads have thicker castings than others, and are capable of being ported to greater sizes. Even if these heads were not designed as performance heads from the factory, if they can be enlarged, outfitted with larger-diameter valves, and have their chambers reworked, they may serve your purposes just fine. On the flip side of this are the heads that may have come on a good factory engine, but are thin-wall castings not capable of being adequately ported or machined for high-performance or heavy-duty use. Research into your particular family of engines, and specifically its cylinder head offerings, will save you both time and money in the long run.

When the term valve angle is used, it defines the angle of the valve stem in relation to the cylinder head deck. If you extended the valve stem past the deck of the head, as shown in the photo, you can see how the valve angle is defined relative to the deck.

To improve airflow, some V-8 designs feature valves that are canted (angled) to one side so the valve moves away from the cylinder wall as it opens. This unshrouds the valve and helps air get into and out of the engine more effectively. It's not hard to spot an engine with canted valves once the valve cover has been removed. The most popular factory canted-valve designs include the big-block Chevy (shown), the Ford Cleveland and Midland, and the Chrysler Hemi.

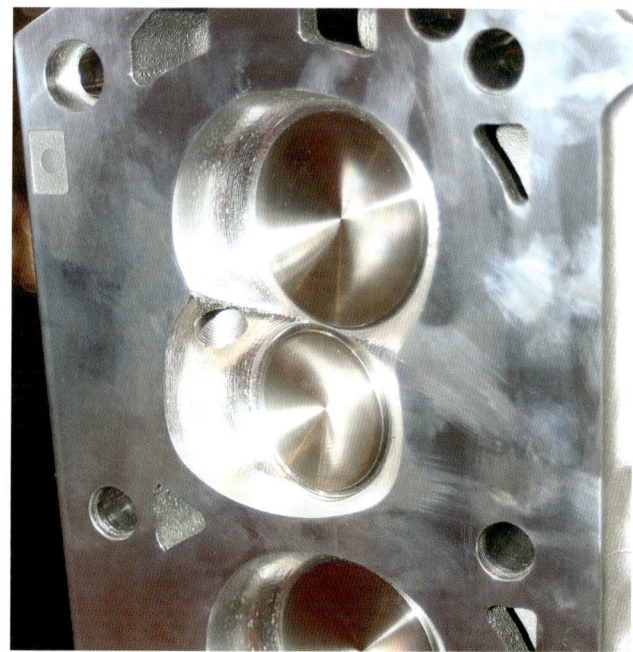

This is Trick Flow's twisted wedge head for the small-block Ford Windsor engine. The engineers at Trick Flow moved the position of the valves slightly, nudging the intake valve a little closer to the intake side of the head, and moving the exhaust valve a little closer to the exhaust port. This allowed them to improve both the intake and exhaust ports for increased flow quality. They also designed the combustion chamber to be more efficient (note its shallow design, unshrouded valves, and CNC finish), and the result is a great bolt-on cylinder head that will make more power than a factory head.

All three valve angle cuts can be seen on the seat here. The widest portion of the cut is the 45-degree valve seat. A gentler 30-degree seat eases the transition into the head, while a 60-degree cut eases the transition into the combustion chamber.

There is a lot to see in this photo. The look into the exhaust port shows a gentle transition from the valve seat into the port itself. The ridges signify a CNC-milled finish on the inside of the combustion chamber, and the spark plug is angled toward the exhaust valve as it should be. When inspecting performance cylinder heads for potential purchase, these are the kinds of things to look for.

A top-quality valve job will enhance flow and permit the valves to seal properly for a long time. While this work used to be done with round stones of various angles, modern machine shops now use a mandrel-located cutter to cut all three (or more) valve seat angles simultaneously. When working with vintage heads, installing modern valve seats for unleaded fuel is a common upgrade, and using a modern tool like this to perform the valve job prepares them for long-term use. The improved flow performance, especially when teamed with a good high-flow, stainless-steel valve, means more power and durability using today's gasoline.

This is another great reason to explore aftermarket cylinder heads. These are designed for performance use right from the start, and typically will require much less work to perform to your expectations. When the cost of reworking a pair of factory iron heads is compared to buying aftermarket heads and fine-tuning them to your particular combination, the aftermarket heads usually end up saving you both time and money.

CYLINDER HEAD DESIGN BASICS

When reviewing cylinder heads, there are some hard and fast basics that inevitably make a difference. While make-specific heads can only offer relatively minor variations, some makes enjoy a broader range of choices in the aftermarket. Also, by learning these differences, should new heads be made available in the future for your chosen make, you'll know what these changes are and how they may benefit you.

The first factor to consider is the intake valve angle. This is a measure of the intake valve's placement in relation to the flat deck surface on the block. In most cases, a lower angle number is better, and we can see this in some of the aftermarket heads available for the small-block Chevy engine. The factory intake valve angle on these heads was 23 degrees, but racing heads have been produced for this engine in 18-degree and even 15-degree variants, each offering better flow than the last. Naturally, special valvetrain gear is required to use these heads, and it's plenty expensive. But, in some racing divisions where these heads are legal and budgets exist to pay for them, they flourish.

Once aftermarket heads became popular, different manufacturers experimented with altering valve angles both in the angle mentioned above and in the side-to-side (canted) direction. By angling the valve sideways, the valve gains clearance to the cylinder wall so that when it opens, it is no

To center the valve seat cutter, a dowel is inserted into the valve guide. The cutting head will be lowered onto this, and then the seat angles can be cut.

longer shrouded by the bore in the block or the combustion chamber. Some factory V-8s have canted valves (like Ford's Cleveland engines and, of course, the Chrysler Hemi) so this is not a new idea. What is new is that some heads now feature a canted angle where the factory never designed one originally, or have added additional angle to existing factory canted-valve designs (like the big-block Chevy). Again, sometimes special valvetrain components are required, and piston-to-valve clearance must be checked carefully. (Pistons are typically notched for valve clearance based on factory valve angles.) The twisted wedge cylinder heads made by Trick Flow are a good example of a cylinder head with altered valve angles that can still be used in a street/strip application without requiring major modifications anywhere else in the engine. I expect this type of design will continue to gain popularity in the future for other engine makes.

Another modification that is gaining popularity is reverse-flow cooling. In this design, coolant is routed to the cylinder heads first, rather than to the block first as was the norm in domestic V-8s for decades. By cooling the heads first, they are capable of withstanding more heat without detonation, because they can shed this heat more effectively. Because this modification typically involves changing the routing of coolant lines outside the cylinder head, it rarely comes under consideration as a cylinder head modification. But I feel that such a modification should be considered at the beginning of an engine's design and development, so I included it here. If you're building a performance engine for street use along

When resizing and reshaping the intake ports, it's wise to use the intake gasket as a template to guide your efforts on both the intake manifold and the cylinder head. Here, we can see how the gasket was used to properly and accurately size the intake ports on this LT1 SBC head. By transferring the same gasket to the intake manifold, we can be confident the intake and cylinder head ports will line up once they are assembled.

Porting your own heads is surely possible, and can add more flow for increased power. However, what you can do at home is limited when compared to what a professional head porter is capable of. By cleaning up the factory castings and squaring up the shape of the ports and chambers, you can't hurt anything. If you've never done this kind of work before, start slowly, work carefully, and don't overdo it and break through a port wall into a water jacket. Making the intake port too large will actually hurt flow velocity at lower rpm levels.

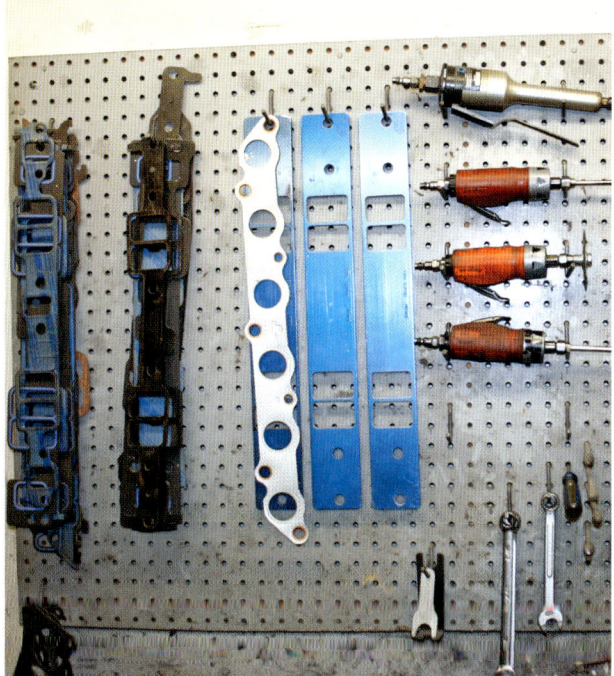

Shown is the porting room of a professional high-performance engine shop. They have all the tools required to work over any cylinder head, regardless of material, and improve its capabilities. More important is the vast experience professional head porters bring to the table. They know where to work, how much material to remove, and most importantly, when to stop grinding!

Using the right tools in the right places is key to effective porting work. Here, a cartridge roll is used on the short-side radius area just behind the intake valve. This is a popular area to work on, as many heads can be improved with some smoothing in this area. If a cutting tool is used, it will remove material quickly. Use of a sanding tool like this one allows gentle shaping to aid the transition from one portion of the port to another.

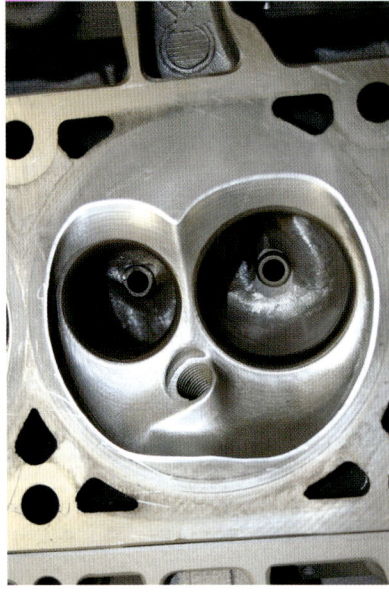

Effective porting work is a thing of beauty, but the power numbers don't lie either. These before-and-after photos show what a talented head porter can do. Notice the revised shape of the chamber to unshroud the chamber, and the shape of the material surrounding the valve guide. The basic heart shape has been shown to be very efficient, and when a modern chamber is shown next to a vintage one, the potential improvements that can be made are obvious. These were done by TPI Specialties (TPIS).

While hand porting is effective, computer-controlled port machining is much more accurate and consistent from port to port. Additionally, the CNC porting programs are based on proven designs, so you're guaranteed a good-flowing head once the procedure is complete.

with occasional time on the track, I recommend looking into plumbing it for reverse-flow cooling. The Chevy LT1 V-8s were engineered this way from the factory, and they proved to be solid performers without the cooling issues typically found in other small-block Chevy engines. I've seen Pontiac, Oldsmobile, and big-block Chevy V-8s, in addition to small-block Chevys, modified for reverse cooling with positive results, so I can recommend this comfortably. The combustion chambers see the greatest amount of heat in V-8s, so it makes sense to send coolant to them first. Keeping detonation at bay is a key to maximizing the performance of any engine, and the addition of reverse cooling really helps engines avoid preignition when the hot spots are adequately cooled.

Engines are commonly described as air pumps, and while this is an oversimplified description, it is relatively accurate. Anything we can do to move more air and fuel into and out of the engine will help it become more efficient and make more power. We already discussed some intake port basics, and while the port is one place where great differences can be made in flow, there are others.

The valve seats are a prime example. When either the intake or exhaust valve closes, it must seal effectively, and the interface of the valve and valve seat is key to this. But then it gets more complicated. On the intake side, air and fuel must flow past the valve and the seat from the port into the combustion chamber. On the exhaust side, spent gases must be pushed out of the valve into the exhaust port. These jobs are very different, yet the designs of the intake and exhaust valves and seats are very similar. It can be said that a doorway works just as well whether one is entering or

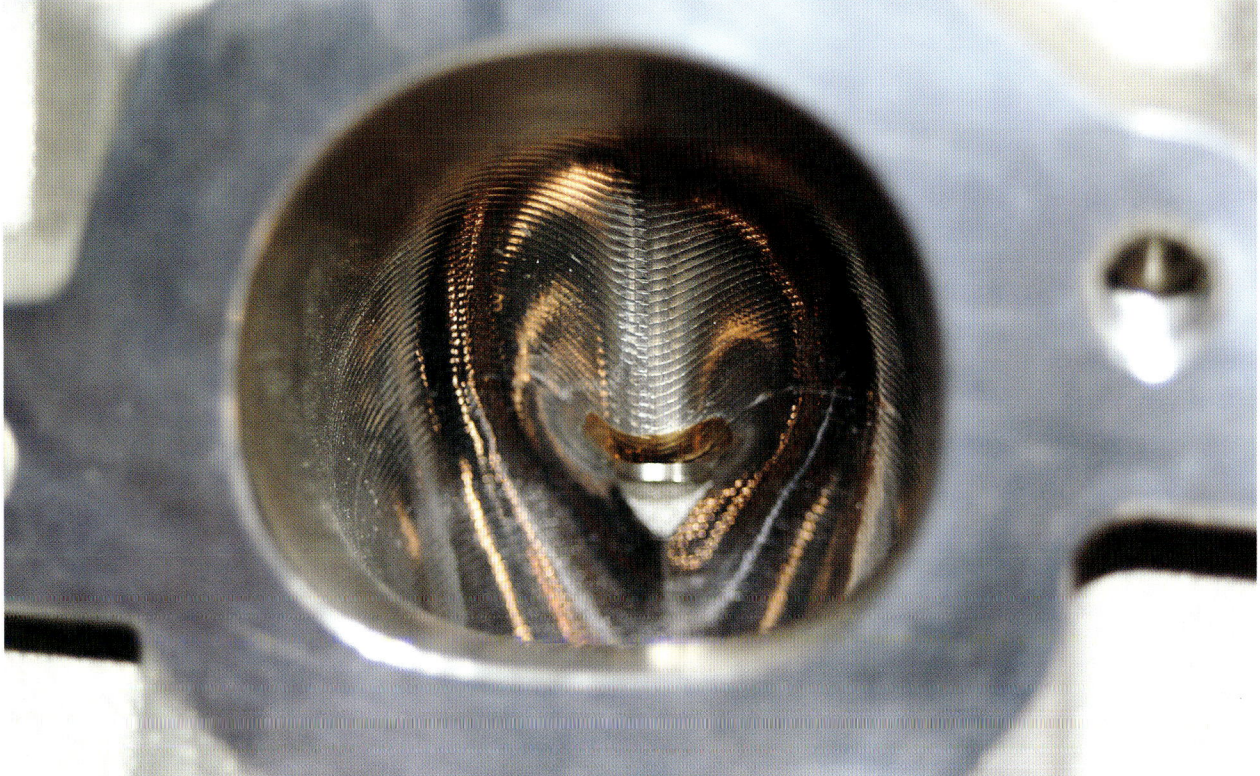

Even the work of a professional head porter is overshadowed by what CNC-programmed machines can do, because the CNC machinery is incredibly consistent from port-to-port and chamber-to-chamber. Also, the CNC programs are based on tons of research to find the best possible port and chamber shapes, and represent measurable performance gains based on these optimal designs and sizes. These are Air Flow Research (AFR) CNC-finished products.

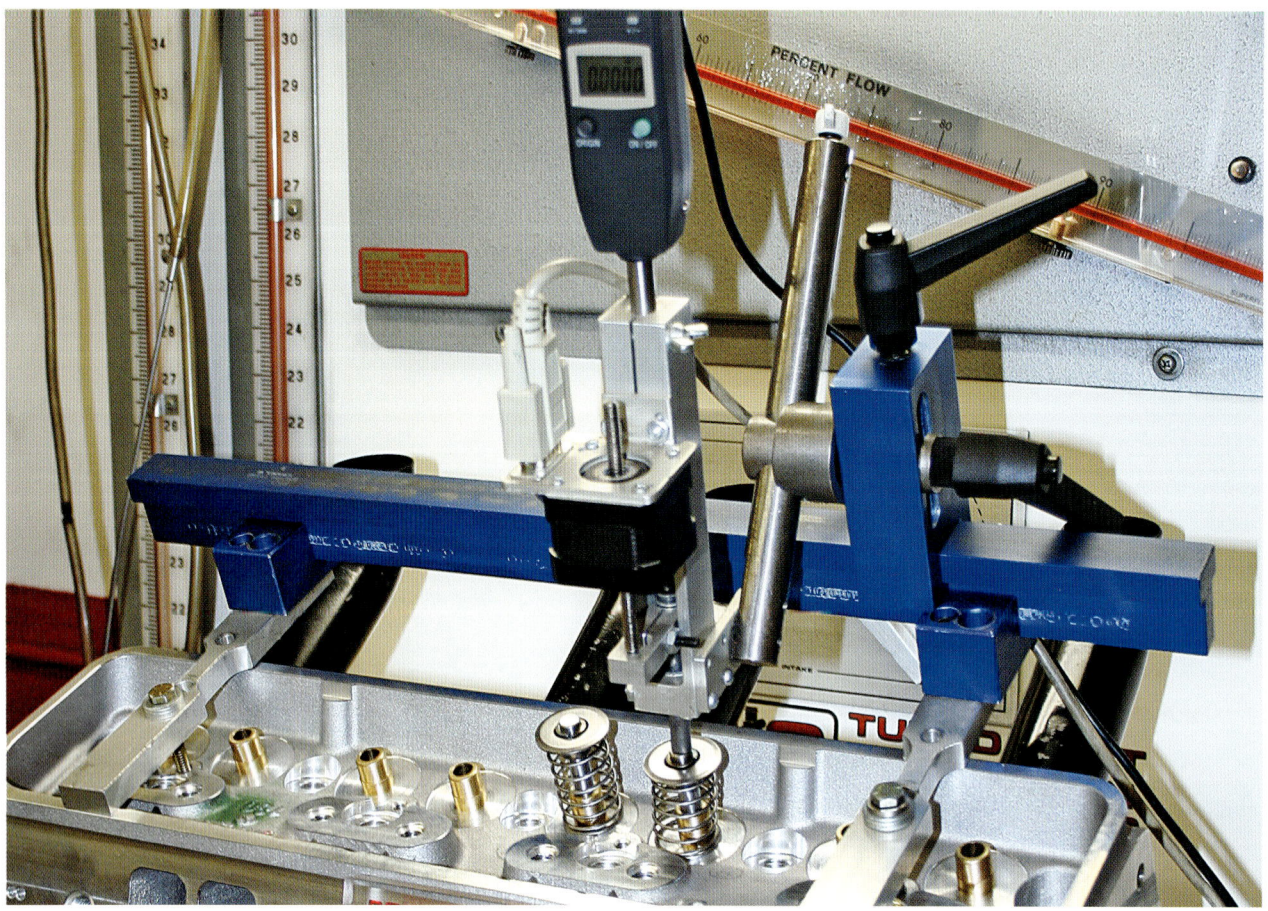

Flow bench testing lets you measure the impact of cylinder head port modifications. It's important for many reasons, including helping you determine what camshaft to choose. Additionally, new technology measures the swirl of the air passing through the port, which can also enhance power. This flow bench has been equipped with both a swirl meter and an automatic test function. The auto test ensures every flow check is done identically to accurately verify improvements and comparisons between different heads.

exiting, but the valves of a V-8 look little like any doorway I've ever passed through.

Most factory heads came with a three-angle valve job, and most performance engines now have four, five, six, or more angles to ease the transition around the valve. The standard three angles are 30, 45, and 60 degrees, with the 45-degree angle meeting the valve. Some have experimented with different angles and found gains in some applications, but the standard angles work just fine in most cases. A radius cutter that blends a radius into the valve seat should also be used. Hardened seats are more important on the hotter exhaust side than on the intake side. I also prefer stainless-steel valves in any performance engine. Aftermarket performance heads typically ship with hardened seats for use with unleaded fuel and stainless valves, but it always pays to ask when ordering to make sure.

The intake port is where most builders find flow improvements, and the key here is to not overdo it. I feel the greatest improvements can be found in simply cleaning up the port walls and ceilings, and leaving the floors alone. Sure, if you've got porting experience and you know what a

particular cylinder head wants, you can do much more than this, but it's impossible to give a generic explanation that would encompass all domestic V-8 heads. Using the intake manifold gasket as a template will help you raise the roof a bit and make the port entry area more consistent from port-to-port. The floor should be left rough to create turbulent airflow there, which will aid in keeping fuel in suspension rather than puddling on the port floor. The short-side radius (the curved area of the port immediately behind the valve seat) is also a place that shows benefits when cleaned up in virtually any head casting. Grinding the valve guide into a smaller size and giving it a teardrop shape in the direction of flow is another proven way to increase performance.

I recommend the porting guide produced by Standard Abrasives for those who have never ported a head before. Standard Abrasives makes a great home porting kit with all the cartridge rolls you'll need to get started, and the advice in their published home porting guide is right on. A good general cleanup will help any as-cast head, and if you're serious about cylinder head porting, it pays to look into the specific needs of the engine family you're working on.

Another new technology becoming popular is computer numerical control (CNC) porting. In this technology, a computer-controlled milling machine finishes the ports and combustion chambers to a preprogrammed finish. The result is identical ports, consistent chamber volumes, and the knowledge the work was done properly to increase the performance of the head. I'm a huge fan of CNC-finished chambers, especially if the company offering them has a solid reputation. Trying to equalize every port and every chamber is virtually impossible by hand, but the CNC mill doesn't know any other way to do it. Consistency is key, and I've found the heads CNC-finished by their manufacturers to be especially good. You get them finished out of the box as new parts, and the heads' flow characteristics are already established. The CNC finish in the chambers is excellent for burn and flow, and the chamber shape has been researched for increased effectiveness. Trying to duplicate such a feat at home, or paying a pro to do so, would be tough. With new CNC-finished heads, you can design and develop your engine around the known capabilities of the head, which has proven to be a saver of both time and money for me.

When reviewing port flow numbers, it's just as important to me to see where the head flows well in addition to the overall flow quantity. If a cylinder head flows great at 0.700-inch lift, but you don't plan on running a camshaft with at least 0.700-inch lift, that flow potential becomes irrelevant. Teaming a cylinder head with a camshaft is key to designing a great engine, and reading a flow chart with some ideas in mind about your cam is wise. Because the head is designed for use on the engine you've chosen, naturally you want the most possible flow to get the best possible performance. Peak flow-bench numbers can tell you about the heads you're looking at, but they are not the final word. The overall flow performance at all valve lift points is critical to the overall design. For street performance cars, the flow at lower valve lift levels (0.300–0.500) contributes more to overall power than the flow at higher lift points (0.600–up), so read the flow numbers carefully and keep your target cam's lift in mind.

The wide range of American V-8s bring with them many different combustion chamber shapes. Some are good, and some are not so good. Today's research has proven that a shallow heart shape is just about the best for good flow and a complete burn, so anything you can do to move toward that ideal is good. Some engines, like this 348/409 Chevy, can't really be shaped in this way, so compromises must be made.

The valves are the doorways into and out of the engine, and upgrading to modern, top-quality valves will surely improve airflow and power potential in vintage engines. These high-flow valves are Manley products, and can be made in a wide range of lengths and diameters with varying stem thicknesses. If you're upgrading the performance of your cylinder heads, the addition of good valves like these is a natural move.

Combustion Chamber Design

The combustion chamber design has a lot to do with how well the head, and therefore the engine, will perform overall. Engineers have been researching the optimal chamber design for decades, and power can be found by adapting some modern discoveries to vintage engines. The keys have been to unshroud the valves so they can breathe more effectively, to keep the chamber shallow, and to control where the burn event happens. Vintage engines with deep, shrouded chambers can be made more efficient by decking the head to make the chamber more shallow, and working with either a flat-top or slightly dished piston to attain the target compression ratio.

Luckily, most vintage V-8 castings are quite thick, and you can shave the deck without sacrificing reliability or capability to seal to the head gasket. Of course, if you remove material from the deck surface of the cylinder head (or the block), the distance between the cam lifter and the rocker will be shortened, requiring a different length pushrod. Naturally, this is something to be discussed with your machinist before ordering him to deck the head, but it is a real possibility in most cases. You may also lay back the walls surrounding the valves if the wall thickness will not be compromised. This will add additional volume to the chamber, so even after the decking procedure, the change to the volume of the chamber will not be as dramatic as it may seem at first. When shaping the chamber, your target should be a heart shape that wraps around the valves. This allows the valves to be unshrouded and gives good flow into and out of the respective ports. Naturally, you need to make sure such modifications arc possible without compromising reliability; if compromise is necessary, so be it. Any improvement you can make, no matter how minor, will help.

Chamber shape affects more than just airflow. The path the burn takes across the chamber is also affected, and this has a direct impact on power production. What we'd like is a small chamber area with a centrally located spark plug. When the plug fires and the flame kernel travels across the chamber, we want it to have the shortest possible route to travel and nothing in its way to restrict the burn event. This is why pop-up, high-compression pistons have lost favor over the years. The portion of the piston that popped up into the chamber would slow the burn event from the spark plug to the far side of the chamber. By moving to a smaller chamber with a flat-top or slightly dished piston, a more efficient burn can be had with the same compression ratio.

With the valves unshrouded and the chamber shallower, you can investigate whether larger-diameter valves can be installed. Depending on the length of the valves you need, it is very possible to install valves designed for another make of engine into yours. Replacing valve guides is standard fare in machine shops, so changing the valve stem diameter should present no problems to your machinist. Typically, the modern- design, high-flow valves you'll want to use will have smaller-diameter stems than the original valves you'll be replacing, so you'll gain some airflow potential there as well. The latest valve designs are great improvements over what was offered from the factories in the 1950s, 1960s, and 1970s, so this upgrade alone should be worth measurable power.

Valve seats are another story, since vintage engines typically have seats made of softer material that used to rely on the lead in leaded gasoline to lubricate them. Modern unleaded gasoline doesn't have this lubricating quality, so valve seats for modern engines are made from harder materials. Installing modern hardened valve seats into vintage engines will allow them to run on unleaded fuel without worry. Any competent machinist should be capable of installing hardened valve seats into vintage cylinder heads.

Optimally, the spark plug should be angled slightly toward the exhaust valve, if possible. Some machine shops can alter plug angles, and if this can be done it should be. Some vintage engines actually angled the plug toward the Intake valve, so even if the plug is made straighter it would be an improvement. By minimizing the distance from the valve to the outer limits of the combustion chamber, a highly efficient burn will result, requiring less ignition timing advance and more effective power production.

The angle of the spark plug is another important design element. Research has shown that the plug should be angled toward the hottest part of the chamber to minimize the effects of preignition. In this AFR head, the plug angle is obvious (as is the beautiful CNC work and exceptional chamber design). This is what a great performance head should look like!

Chapter 6
The Intake Tract

The intake tract is defined as everything that handles air up to the intake valve. Naturally, the intake manifold is a big portion of this tract, and how it directs the engine's incoming air (or air/fuel mixture) to the intake valve will have a dramatic impact on the engine's performance. The intake manifold's plenum volume; its runner volume, length, and shape; and the angles it must direct air and fuel around are all critical. Whether the fuel delivery system is carbureted or fuel injected makes a difference too. As the engine designer, your task is to choose the best possible intake manifold to suit your particular engine's purpose and rpm range. The purpose and target rpm range of the engine you're developing are critical to know ahead of time, and compromise is key. In a naturally aspirated engine, there will be a bit of compromise simply because the engine will be more efficient at some rpm levels than others.

Let's discuss carbureted applications first. There are a few basic rules we should understand in this application, as both air and fuel are drawn into the intake manifold by negative pressure (vacuum) from each individual cylinder. This means the fuel, which is much heavier than the air it's traveling with, will constantly try to fall out of suspension as the mixture heads toward each intake valve. Keeping the fuel in suspension is aided by rpm, since the air/fuel column gains velocity with rpm, and the fuel has less opportunity to fall out.

Volumetric efficiency is a common term that defines the engine's efficiency in being able to draw air/fuel charge in and move exhaust charge out. It is measured as a percentage based on how much air and fuel mass enters the engine compared to its actual static piston displacement. Typical production engines reach more than 80 percent volumetric efficiency at peak effectiveness, and with forced induction (like a supercharger), levels of more than 100 percent are attainable. Exceeding a volumetric efficiency of 100 percent means the engine actually moves more measured cubic inches of air/fuel through it than it displaces, which is a lofty goal. The more efficient an engine is in this regard, the more power it can potentially make.

The two basic intake manifold options for domestic V-8s include dual-plane and single-plane designs. They are quite different in appearance and not hard to spot. The dual-plane design (left) separates the incoming charge to separate plenums, while the single-plane feeds all eight cylinders from a common plenum. Shown are Edelbrock products for the 348/409 (dual-plane-left) and the big-block Chrysler (single-plane). Edelbrock offers both styles for a wide range of domestic V-8s of all vintages.

Research has shown that adding a notch in the divider wall directly beneath the carburetor, between the left and right sides of a dual plane intake, is worth some power. Now even brand new dual plane manifolds, like this Bow Tie intake from GM Performance Parts, come with the notch already cut.

The improvements we make in getting air into and out of the engine contribute to improving its volumetric efficiency, especially at higher rpm levels where such changes can have a greater effect.

Both volumetric efficiency and power can be increased considerably by taking advantage of the natural dynamic effects that occur during the intake filling cycle.

Engine designers have long known about the resonance effect in an engine's intake manifold. When the intake valve opens, the air inside is sucked into the cylinder. When the valve closes, air stuck inside the intake tract bounces against the closed valve and rebounds backward. After traveling the length of the tract, it hits the front of the intake manifold and heads toward the intake valve again. The rpm at which this pulse occurs varies with the length of the intake port, and in any case only occurs during a narrow rpm band. A clever engine builder can tune the length of the intake tract to precisely time the arrival of an incoming pulse with the valve opening. This natural supercharging effect packs slightly more intake air into the cylinder, at that rpm, increasing power.

With changing rpm levels, the amount of air and fuel moving through the intake manifold is changing constantly as rpm changes. This means the optimal efficiency point for a given intake manifold design is relatively stable, and as the rpm moves around this optimal efficiency point, efficiency ebbs and flows.

The design of the intake manifold is a major determining factor on where this point will be, so it's no surprise a wide range of manifold designs exist. The two basic design types

used with carburetors are single-plane and dual-plane. A single-plane intake manifold shares the plenum among all eight cylinders, where a dual-plane intake manifold splits the plenum into two separate chambers. In a dual-plane intake, the typical design separates the plenums based on the firing order. Using a typical 1-8-4-3-6-5-7-2 firing order, cylinders 1, 4, 6, and 7 will share one plenum, while cylinders 8, 3, 5, and 2 will share the other plenum. This way, as the engine runs, it pulls equally from each side of the intake manifold, and the rate at which air and fuel are being delivered to each cylinder is basically equalized. Naturally, due to space limitations, there are differences in the length of each individual runner, and the effect on performance is an acceptable compromise.

A single-plane intake, by comparison, forces all the cylinders to draw from the same plenum, allowing a more direct path to the intake port from the air/fuel source at the carburetor. Research on early single- and dual-plane intake manifolds showed the dual-plane designs offered better low rpm performance, while single-plane designs made better peak power at higher rpm levels. For many years, this was the rule, and for the most part, it was true.

Today's best dual-plane intake manifolds are capable of performing well up to the 6,000–6,500-rpm level, while today's best single-plane intake manifolds begin making good power as low as 3,000–3,500 rpm. It's been my experience that a large-displacement engine typically makes a bunch of torque at low rpm levels regardless of the intake manifold design. Once you get past 460 cubic inches or so in a big-block, and around 360 cubes in a small-block, low rpm torque is not hard to find. In fact, many

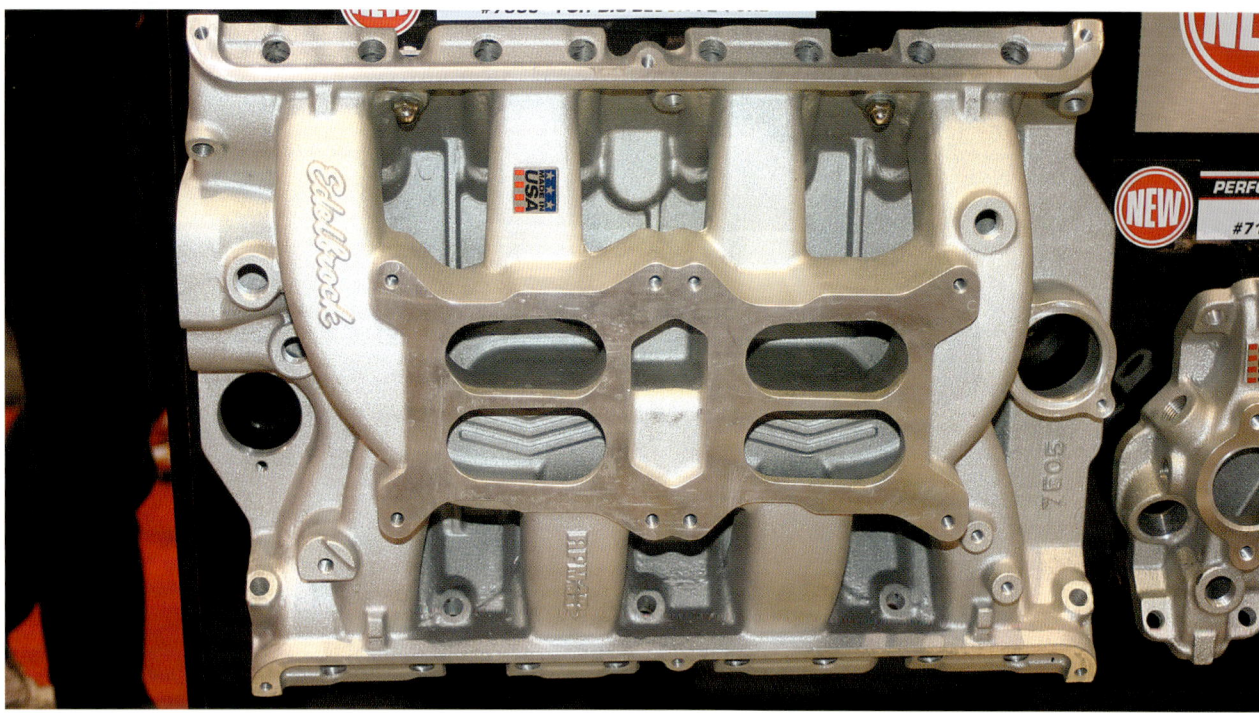

This new dual four-barrel intake for the classic FE Ford V-8 delivers good performance (thanks to its dual-plane design), has vintage high-performance looks, and is low-profile enough to clear most hoods.

This Mopar Cross Ram intake is a factory design, and was developed to make big power at wide-open throttle at the drag strip. It does this well, and has been atop many record-setting Chrysler products since its introduction in the 1960s right up to today. In the Super Stock classes, it proved to be a killer. On the street, it's more of a challenge to tune, but it can be done with the right cam and proper carb setup.

You can fine-tune your plenum volume with carb spacers. They come in many styles and are made from many different materials, including aluminum, phenolic, and composite. Shown are a four-hole spacer from BG Fuel Systems (left), and a selection of aluminum and phenolic spacers of varying thicknesses.

The oxygen sensor, placed in the collector of the exhaust header, measures the oxygen in the exhaust stream to determine if the engine is running rich or lean. While oxygen sensor technology advanced because of its importance in electronic fuel injection systems, new stand-alone oxygen sensor systems are now being used to help carbureted engine owners fine-tune their air/fuel mixtures as well.

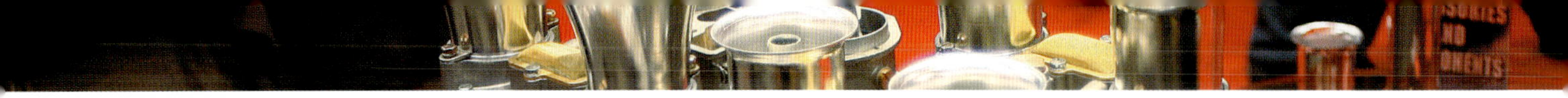

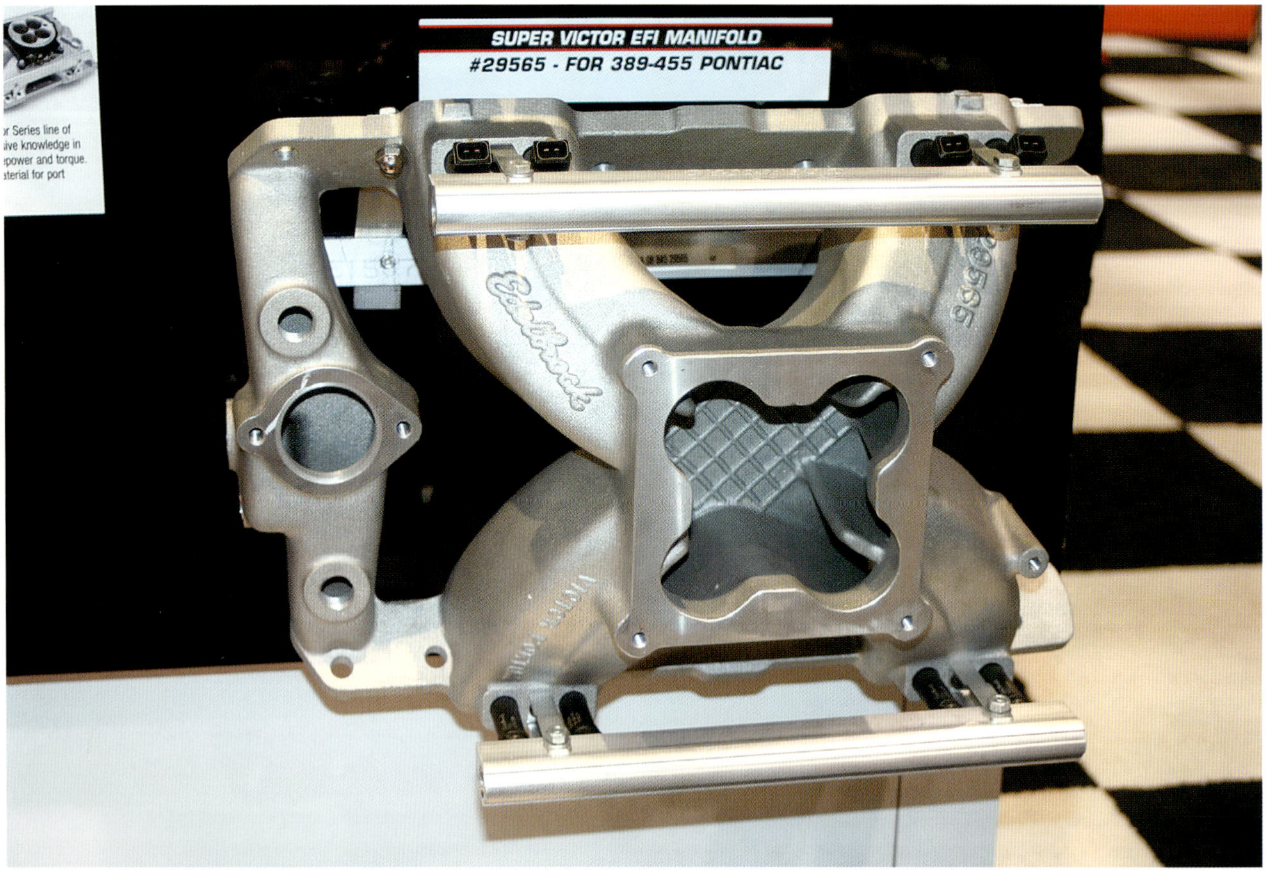

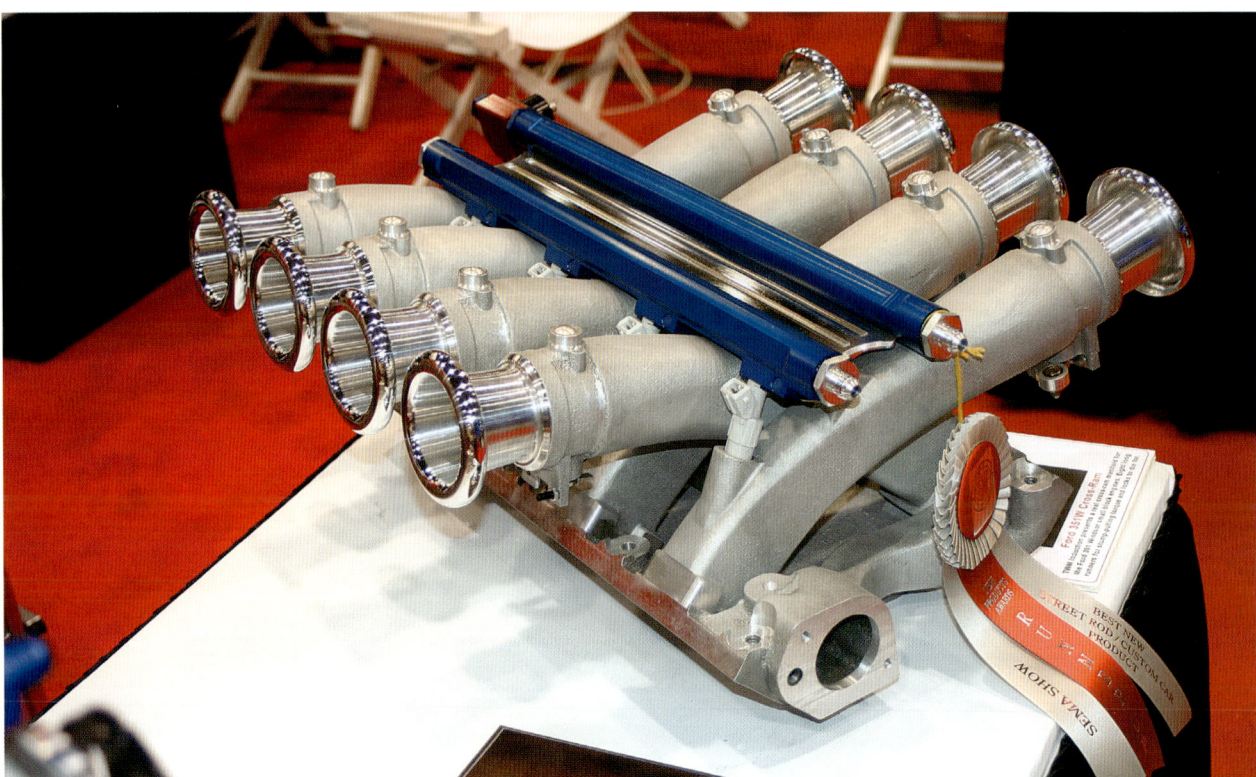

SUPER VICTOR EFI MANIFOLD
#29565 - FOR 389-455 PONTIAC

BEST NEW
STREET ROD/CUSTOM CAR
PRODUCT
SEMA SHOW

EFI intakes come in a wide range of styles, from upgraded single-plane carb-style intakes (like the Edelbrock unit on the top, for traditional Pontiac V-8s) to wild-looking, eight-stack setups (bottom). You can also upgrade almost any carb-style intake by having it plumbed and adding the fuel injector bungs.

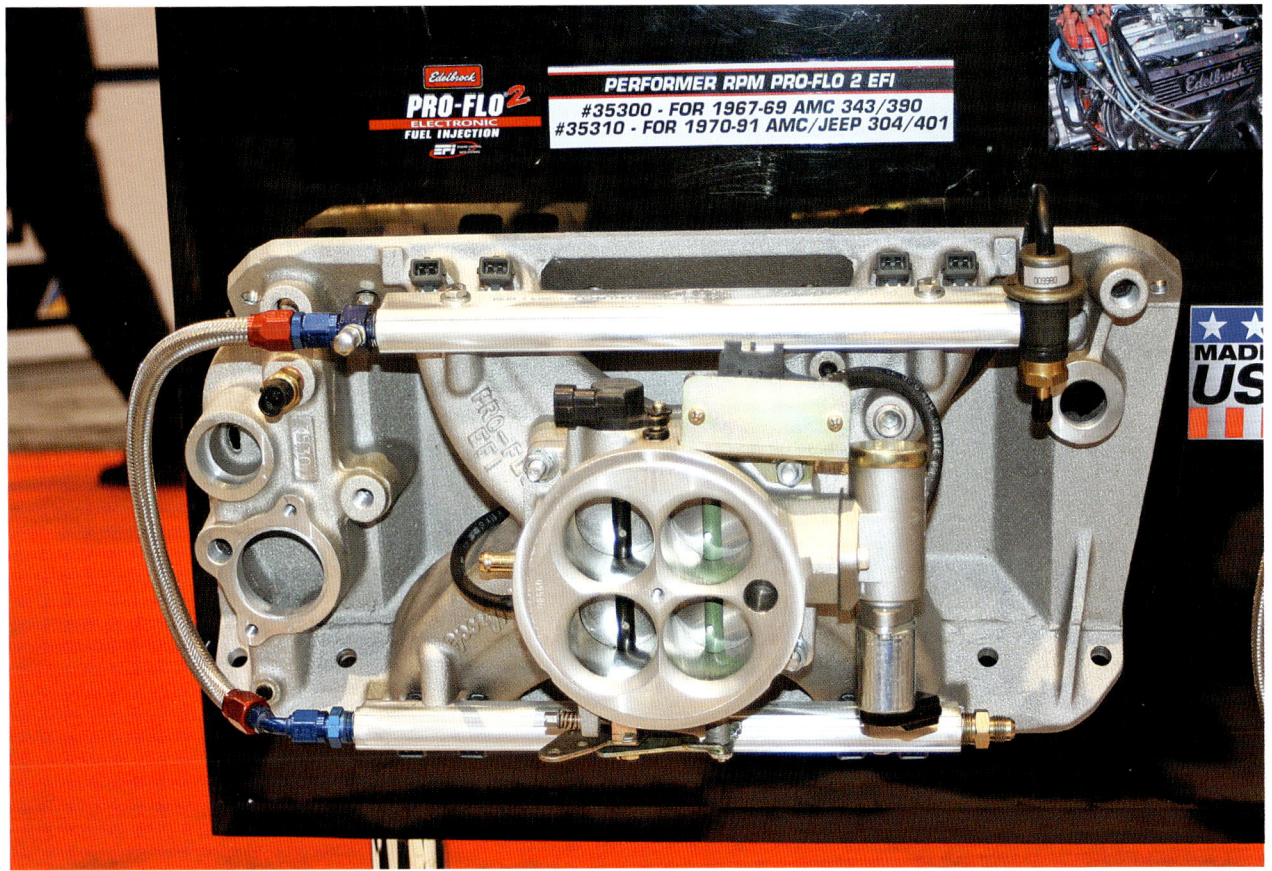

Complete EFI upgrades are becoming more common, and less-popular engines are finally being given some overdue attention. Here's a complete EFI setup for an AMC V-8, which may not sound like a wise marketing move until you consider how many Jeeps were sold with AMC V-8s! EFI helps the 4x4 enthusiasts plenty because the fuel system is no longer affected by the angle of the vehicle. Carburetors, with their float bowls and the jets that feed from them, are affected by vehicle angle and the bumpy ride common in off-road applications. EFI cures all these issues.

racers advance their camshafts a bit to kill some low rpm torque and gain traction off the line. For these reasons, I like to run a single-plane intake manifold on a large-displacement engine destined for street and strip use. By maximizing performance at higher rpm levels, these engines are allowed to really flex their muscles once the car has the best possible chance of putting that power to the ground. This is especially true when these big-displacement engines are installed in lighter (less than 3,500 pounds) vehicles with good rear suspensions and proper tires.

On the flip side of this, smaller-displacement engines need all the help they can get at lower rpm levels, and a modern dual-plane intake can really wake them up. Because the modern dual-plane designs typically don't fall off until they're past 6,000 rpm, it's not much of a sacrifice anyway, and if the engine is well designed for midrange power (3,000–5,000 rpm), it should pull like crazy on both the street and the strip. Of course, the vehicle weight and suspension capabilities are once again a concern, but with a smaller-displacement engine (less than 450 cubic inches for big-blocks, and less than 360 cubic inches for small-blocks), odds are good you'll want all the low-end power you can get.

Modifying intake manifolds typically involves grinding the manifold ports to better align with their corresponding cylinder head intake ports. This can be done using an intake manifold gasket as a template and working carefully to keep the ports identical in size. This is called port matching for obvious reasons, and has been shown to improve power throughout the rpm range when accomplished correctly. Using a gasket as a template was discussed in the last chapter, and the same lessons apply to the intake manifold ports.

An alternative trick I've found that works well on street engines with carburetors is to leave the intake port a bit smaller than the corresponding cylinder head port. The air and fuel is kept in suspension as the mixture flows over the step between the manifold and cylinder head. This won't work in port-injected EFI engines where no fuel is in suspension, or on high-rpm race engines where the air and fuel are moving so fast there's little chance for the fuel to fall out of suspension. In these cases, matching the intake port to the cylinder head port is the best course of action.

Once installed, the wide variety of EFI intake designs make for some really unique looking engines. The great look of injector stacks that became popular on mechanical injection systems in the 1960s can now be civilized for street use.

Other modifications typically found on intake manifolds include porting for increased flow potential from the inlet down to each port, and (on single-plane designs) the extension or reduction of the walls separating the intake ports where they flow into the common plenum under the carburetor mount surface. On dual-plane intakes, builders have sometimes found power by cutting a notch into the wall that divides the left-hand side of the intake from the right-hand side under the carburetor mount surface. This has proven so effective that many dual-plane intake manifolds are now being cast with the notch already in place.

These modifications may seem rather minor, but there are no moving parts in the intake manifold. If it is designed and sized correctly for the application, there's not much more to do.

Technically, long-ram intakes, like tunnel rams or cross rams, are single-plane intake manifolds and can be treated as such. Their longer runner lengths make a difference in how they will respond, and these benefits typically show up at high-rpm levels. When designed for use at WOT at the drag strip, tunnel rams and cross rams can deliver

air and fuel effectively when teamed with other high-rpm equipment (like a suitable camshaft). The larger plenum area and longer runners both contribute to these effects, and while it may not seem possible, some have been used successfully on street engines.

I have found success in running tunnel rams and cross rams on the street by using a pair of small carburetors (typically 500 cfm or less) and relatively low-lift, short-duration camshafts. The small cams offer plenty of vacuum, and the small carburetors are able to remain responsive even when they are mounted a relatively long distance from the intake valve. I have had to run larger accelerator pumps on these setups to get crisp throttle response, but otherwise they've proven to be stout performers on the street. At the track, the small cams limit the peak performance potential of the engine combinations, but the owners of the cars wanted to maintain street driving capabilities, so this compromise worked well.

The use of spacer plates sandwiched between the carburetor base and the intake manifold has become popular lately. These serve several purposes, but the basic benefit is increasing the plenum volume of the intake. This will make a difference in the performance of the engine, and my research has shown that this typically results in a bump in midrange power and torque, but every engine is different and will respond differently. Aluminum carburetor spacers also move the carburetor away from the intake manifold a bit, and can aid in keeping engine heat away from the carb.

There are three basic designs of carb spacer: open, four-hole, and tapered four-hole to open. Personally, I like the tapered four-hole to open spacers best because I've seen them deliver more positive results than the other two types, but as every combination is different, this is not a hard and fast rule. If you can, test all three types and see which works best for you. Most good dyno shops have these available for testing, and it's an easy enough modification to make, so testing is a good move. A spacer will, naturally, raise the carburetor up, so if hood clearance is tight, check carefully before closing the hood.

While carburetors are the traditional standard for feeding fuel to big vintage V-8s, the advent of EFI technology has been adapted well to these engines. EFI is more accurate and flexible than a carburetor ever could be, and once tuned, provides a wide range of benefits that justify its expense. Easy starting (both cold and hot), the promise of proper air/fuel ratio throughout the entire rpm range and improved fuel economy are typical benefits EFI can bring.

You'll notice I did not mention horsepower or torque, since a perfectly adjusted carburetor will make as much power as perfectly adjusted EFI will. Since EFI relies on exhaust oxygen sensor readings and other data such as air temperature and mass readings to tune itself, the odds are typically better that EFI will be in proper tune at any given time. Not all carbs can do that, unless the owner is capable of tuning and maintaining the carburetor properly.

EFI systems can deliver fuel either through a throttle body or through injectors placed directly in the ports. Either of these systems can bring reliable performance, but direct port injection systems are preferred by enthusiasts. When the fuel injectors are mounted in the intake ports directly ahead of the intake valve, fuel distribution is not an issue. Once this is done, the intake now moves only air, meaning fuel cannot fall out of suspension or puddle up. The intake manifold design is much less of a concern, and many port fuel setups rely on free flowing, single-plane intake manifolds. The carb is replaced by a simple air valve (often called a throttle valve or throttle body), and more power is typically the result. The power gains can be credited to proper fuel distribution, which is a great benefit of port-injected EFI systems.

Throttle body EFI systems replace the carb with an integrated injector system. These throttle bodies dispense fuel from the same place a carb does, so the concerns over wet flow characteristics and fuel falling out of suspension still exist. Still, the computer adds or subtracts fuel as required to maintain the proper air/fuel ratio, and many enthusiasts like the look of the carblike throttle body on their vintage engines. Naturally, the plumbing is much simpler than port fuel injection systems, and the up-front buy-in expenses associated with throttle body injection are typically lower than for comparable port injection systems.

Still, carburetors are the most popular and budget-conscious way of delivering air and fuel to vintage engines, and selecting the proper carb for your application is a critical move. Choosing a specific brand is the first big step. It's been my experience over the years that the carburetor brand you know how best to adjust is the best carb for you. If you've worked with Holley brand carbs and know how to adjust them, but you've never worked with an Edelbrock or Rochester Quadrajet before, you should stick with the Holley or be willing to learn how to fine-tune one of the other brands.

The most common carb makes include the Holley, Demon, Edelbrock, and Rochester Quadrajet. Additionally, vintage carbs like Carters and Strombergs also have a strong following. All carbs have some universal similarities, and while the method of adjustment and tuning varies from make to make, all carbs have a basic series of circuits to deliver fuel at varying engine rpm levels. These include the idle circuit, midrange circuit, and high-rpm circuit. Each will also have an accelerator pump circuit of some kind to give an extra shot of fuel when the gas pedal is depressed. How each type of carb delivers this fuel and transitions from circuit to circuit also varies, and this is where learning the specifics of your carb of choice really come into play. There are ways to adjust each of these circuits, and while many carbs are really close to being good right out of the box, almost all require fine-tuning to be as perfect as possible.

If you've got a vintage V-8 you'd like to build up for increased power, some parts may be hard to find. While it's great that so many new intake manifolds are being cast for vintage engines, the best place to find some interesting options may be the swap meet. This dealer specializes in rare multicarb intakes, which are becoming very popular. They are expensive, but this dealer typically cleans and inspects the intakes he sells, which helps when the price tag is steep.

A look around the rest of the dealer's booth shows a wide range of carburetor and injector setups. Because intakes contain no moving parts, they typically don't wear out like so many other engine parts. The intake setup is what people see when you open the hood (if you even have a hood), so a cool intake really makes a statement. If you can make these vintage speed parts really work well, the fun cannot be measured. Building an engine around the capabilities of a good-looking vintage intake is a good way to do it.

One last look in the booth shows intakes for straight sixes and even straight eights, in addition to vintage Roots blowers and more. I know this book is about V-8s, but if you're an engine guy like me, you love to see this stuff!

Modern carbs offer options not available in the past. A good example is the Demon RS (removable sleeve) carb. By changing the sleeve inserts, the cfm potential of the carb can be altered from 675 cfm to 775 or 825 cfm. Demon offers a few different variations on this carb too. The green sleeves installed show this one is set up to flow 675 cfm, which is perfect for the hot 383-ci small-block Chevy underneath it.

The Holley four-barrel carburetor can be easily identified with its distinctive fuel bowl shape on the front and rear. Holley carbs were embraced by hot rodders because of their reliability, simplicity, and tuning capabilities. They remain a very popular choice on modified V-8s today.

A huge number of GM V-8s were equipped with Rochester Quadrajet carburetors. While they lack the simplicity of design the Holleys were known for, a Quadrajet can be a fine high-performance carburetor with proper tuning. You can easily see the size difference between the small primary barrels and the huge secondary barrels. This spread-bore design offered decent fuel economy and solid performance potential. Both of these attributes make it a great choice for those who can tune them effectively, or are willing to learn.

Weber two-barrel carbs, like these four atop a small-block Chevy, offer a carb barrel for each individual cylinder. This aids tuning, as each cylinder can be isolated and tuned individually. Webers are known for needing regular tuning adjustments, but when they are perfectly tuned, they work wonderfully. These visually impressive setups are quite expensive when compared to a single four-barrel arrangement, but they get a lot more attention too.

These are the removable sleeves used in the Demon RS carb. They are not difficult to change, and offer size flexibility and range not available in other mainstream carburetors. They are color coded for easy reference.

If you are unfamiliar with carburetors and have no experience tuning them, I have to recommend one of the entry-level carbs sold by these manufacturers. The Edelbrock carbs in particular have a great reputation for being ready to run out of the box with a minimum need for adjustment, beyond the most basic idle settings. How these adjustments are made is covered thoroughly in the paperwork that comes with the carb, and even if you've never touched one before, you can get an engine running if you follow the basic steps outlined in the directions.

When setting up a new carb, begin by making sure the throttle linkage is adjusted so the carburetor's throttle plates open completely when the gas pedal is completely depressed. Then, adjust the idle screw to maintain a steady running engine (typically between 700 and 900 rpm in street engines). Next, the idle air/fuel screws can be adjusted based on the manufacturer's instructions. These basic adjustments should get the car running well enough to be driven, and then more advanced adjustments (like jetting) can be tuned in. Again, the manufacturer's instructions should be referred to here.

In some cases (like Holleys and Demons), you should also check the float bowl fuel level while setting up a new carb. While the fuel bowl level is critical with all carburetors,

Here's another model of Demon carb, this one with vacuum secondaries and an electric choke. While it lacks the removable sleeves, it still offers a wide range of adjustability for fine-tuning to a specific engine combination. This is a 750-cfm model.

Looking down into the top of the carb, we can see the air bleeds for the four-corner idle circuit and high-speed circuit on the edge of the carb venturis of all four barrels. By feeding the idle from all four corners, distribution is improved. The air bleed sizes can be increased or decreased to fine-tune the air/fuel ratio in these two critical circuits. Naturally, you'd have to acquire air/fuel data (typically from an exhaust-mounted oxygen sensor) to accomplish this tuning.

The popularity of vintage engines is prompting the development of interesting new speed parts, like this four-Weber setup for a Buick Nailhead. While multicarb setups require experience to tune properly, the decision to invest in a setup like this should prompt the owner to become educated on the tuning requirements. Many enthusiasts are hungry to learn more about performance engines. Designing and building an early Nailhead around the capabilities of this impressive carburetion setup would make learning the carb tuning requirements a joy for many.

It's an advantage to expose as much of the air filter element as possible to minimize any restriction of air flow to the inlet system. In the case of this tri-power-equipped engine, a wide triangular filter element was engineered to feed the three carbs, but it is surprisingly thin. The cover of the air filter element was removed for this photo. Minimal hood clearance in this application limited the potential height of the air filter, so the wide triangular shape was a solution to increasing the filter element surface area to limit restriction.

While it's possible to mill down the top of a Holley carb to smooth out the incoming airflow, adding a Stub Stack is a budget way to accomplish the same thing. These have been proven to aid power in most cases, and it's an easy mod to make.

Holleys and Demons can easily be checked and adjusted externally, while Carter/Edelbrock and Rochester carburetors require partial disassembly to adjust.

Determining the right airflow requirement for your engine ensures your carb is capable of feeding your engine properly. Airflow is typically measured in cfm, and this is how you'll find carburetors listed. If you choose too small a carb, you'll leave some power behind, while too large of a carb will leave you with drivability issues. The accepted formula for determining the cfm requirements for any engine is based on the cubic-inch displacement of the engine, the engine rpm level the engine will be pushed to, and the efficiency the engine is capable of.

The formula looks like this:

$$cfm = rpm \times displacement/3456$$

In practice, let's determine the cfm requirements of a typical small-block Chevy at 350 cubic inches with a 7,000 rpm peak. Rpm (7,000) x displacement (350) / 3,456 equals 709, so this engine needs 709 cfm if it's 100 percent efficient. We then determine the engine's efficiency level, and apply it to this figure as a multiplier.

I like K&Ns filtered air cleaner lids because they ensure plenty of air is available to the carb. While a typical 14x3-inch round filter should be able to provide plenty of air to a typical street-based four-barrel carb, the filtered lid add-on guarantees it.

Typical stock engines are only about 75 percent efficient at peak power, and about 80 percent efficient at maximum torque. As we improve the performance of our engines, these numbers can increase to around 80 percent at peak power and 85 percent at peak torque. Full-on racing engines bump these figures to around 90 percent at peak horsepower and 95 percent at peak torque. Supercharged engines can improve these figures to 100 percent or better. How is it possible to have an engine with more than 100 percent efficiency? When a supercharger is used, more air and fuel than is theoretically possible to be moved through the engine is forced into it, so the resulting benefit is more than 100 percent efficiency.

The efficiency multiplier is simply figured into the cfm figure. For instance, using our 7,000-rpm 350 once again, we'll assume a really good 85-percent efficiency level. We multiply our 709 cfm figure by 0.85, and end up with 602 cfm. So a 600-cfm carb should be fully capable of feeding this 7,000-rpm 350. This may seem like a small carb, but a well-tuned 600 will provide our example engine with crisp throttle response and should be able to bring it right to redline without protest. A 650-cfm carb could be tuned to provide good performance as well, but a 700- or 750-cfm carb may prove to be too much. Naturally, carbs run at many other throttle positions besides wide open, and effective tuning at all these points is what separates a good engine from a great one. This is why learning to tune your carb of choice is such a big deal.

The best advice I can offer is to purchase a good book on tuning your carb of choice, and study it thoroughly before even beginning to play with your carb adjustments. Then, once you've educated yourself, use the book as a guide to fine-tuning your carb. Make one change at a time, and drive the car each time to determine the effects of the change. Don't hesitate to read your spark plugs after every change and keep a notebook that reflects the impact of each change you make. (You don't have to read all the plugs, but at least check one on each side of the engine to ensure you're on the right track.) This is how you learn, and reviewing your past work will make answering future questions easier.

Tuning the accelerator pump shot, when the secondary throttle plates open (on four-barrel carbs), and the overall jetting at WOT are the major issues, but don't overlook idle performance. Idling is one of the hardest things an engine must do, and street engines spend much more time at idle than they do at WOT. Having a proper, steady idle can be a real challenge in some cases, especially when high-lift, long-duration cams are used. Additionally, many street engines rely on vacuum to operate other accessories, like power brakes. Easy starting, a lack of vacuum leaks, and endless issue-free idling truly civilize a high-performance engine for street use. Time invested in idle tuning is always time well spent.

Finally, you must filter the air entering your engine to ensure no dirt gets in and scores your cylinder walls or damages your bearing surfaces. Choosing a good-quality air cleaner is easier now than it's ever been, thanks to advances in oil-impregnated filter technologies. The wide range of traditional round air filters and the newer cone-shaped filters now offered trap more dirt and flow more air than their paper predecessors. Additionally, many of the new-style manufacturers are willing and able to create custom filter shapes to meet your specifications, should you need something unusual. These new-style filters can be cleaned, reoiled, and reused over and over again, so the investment is worthwhile and the performance is exceptional.

Chapter 7
The Exhaust system

When discussing exhaust system design for high-performance V-8 engines, tubular headers are almost a given. However, factory exhaust manifolds can still be a functional option for those who cannot run headers due to space or budget considerations, and they can be made to flow better with some careful porting work. We've seen how to use gaskets as porting templates on both the cylinder heads and the intake manifold, and the same holds true for exhaust manifolds. Opening up the mating port surfaces of the exhaust manifold can improve the performance of stock manifolds. While they may not perform on a level comparable with tubular exhaust headers, they can be made substantially better.

Tubular headers have been proven to add power and torque, in addition to greater fuel economy, when compared to factory manifolds. Once you've decided to run them, the only question that needs to be asked and answered is which headers are best for your particular application. I have wrestled with this question many times over the years, and I've concluded that the most important factor is fit.

While primary pipe diameter and length are important considerations, they are much more critical in racing applications than they are in street applications. Typically, smaller-diameter headers offer better fit based simply on space limitations. Fit is such a big concern in the real world that I must suggest you purchase smaller-diameter headers that fit well versus larger-diameter headers that may offer a bit more power at higher rpm levels.

If you must run exhaust manifolds (as opposed to tubular headers), you can still improve the performance with some porting work. It's easy to see where this manifold could benefit from some effort, especially near the top of each port.

For a true street car, header fit is a primary consideration. These headers (by Hedman) aren't full-length, as the collector is pretty high up on the pipe and the primary tubes do not extend under the car. This costs a few horsepower, but adds necessary ground clearance under this lowered car. Note how these headers offer clearance to suspension and steering components, and good access to the spark plugs. Also, these headers are coated for both appearance and improved performance.

If you've got the ground clearance, long tube headers typically deliver more horsepower and torque at low- to midrange rpm levels. Top-quality headers are now available for most V-8 powered rides, right up to smog-legal versions for late-model cars and trucks. Headers offer true bolt-on power gains and even increased fuel economy in some applications.

Another locking header bolt option is this insert-style fastener from Percy's Performance. They rely on splits in the hollow fastener, which are forced outward to lock it in place when the center socket screw is tightened. If you have sufficient access to the fastener locations, these work great too.

Street cars will spend most of their lives woring in the low rpm ranges where smaller-diameter headers deliver better performance than larger pipes anyway, and not having to dent and/or dimple the pipes to fitaround suspension components, steering columns, or clutch linkages is wise. Having easy access to the spark plugs is also a major consideration, because removing the plugs for tuning reasons and for regular checking or replacement is a big part of owning a high-performance V-8. Also, consider that the header collectors will probably be the lowest part of the car, and if your car is lowered, these will be even more subject to damage. A smaller-diameter set of headers can ptentially be tucked up tighter than a larger-diameter set.

If you're fortunate enough to have several different header options, opt for a tri-Y design if you can. While a typical four-into-one design is good, research has shown that tri-Y headers deliver a broader power band than comparable four-into-one styles. Rather than plumbing all four primary pipes into a common collector, tri-Y headers team the four primary pipes into two secondary pipes, and then into a collector. The tri-Y design is more complex and typically more expensive, so whether or not it is worth the extra expense is up to you. Again, fit is the big thing, but in many cases, the superior tri-Y designs can also fit easily. They may require more investment, but they are something to shoot for if you can.

Header bolts are notorious for coming loose. This is caused by the heat cycles they see (the bolt holes get bigger when they heat up) and the vibration inherent in a heavily cammed V-8. To prevent the bolts from coming loose, invest in a set of locking bolts, like these from Stage 8. Once installed, you can see how the locking mechanism works. The lock limits how much the bolt can turn, and it cannot work its way out with the locks properly installed. Access can be an issue, but with a little effort, they can be installed pretty much anywhere.

The Percy's fastener also installs easily, if you have adequate access to accomplish it. It can be tough or impossible to get the hex head tool in place to tighten the inserted bolt in some cases, but with good access they are a breeze to work with and will not loosen at all.

There has been a lot of research invested into exhaust systems aft of the primary pipes lately, and a new generation of header collector designs have proven to make power. Many of these new designs can be adapted to existing headers, and if the expense of the modification can be justified, they are a good move. By necking down the collector just aft of the primary pipes and extending the goilet (the point where the pipes join together), performance can be improved.

Additionally, routing both sides of the exhaust system together with an X-pipe has also shown to be worth power. Based on the firing order of traditional V-8s, each side of the engine works independently on each revolution of the engine. By joining the exhaust systems with the X-pipe, the odd firing cycle feeds into the opposite side of the engine and helps pull exhaust gases out. This phenomenon, known as scavenging, makes the entire system more efficient.

When you look at the firing order of a V-8, you'll see that the cylinders don't fire from side-to-side in order. Rather, they fire on one side, then the other, then two cylinders will fire consecutively on one side, then it jumps back and forth again, and two cylinders on the other side of the engine will fire consecutively.

For example, a typical GM firing order pairs the odd-numbered cylinders on one side (1, 3, 5, 7) while the even numbered cylinders are on the other side (2, 4, 6, 8).

Look closely at this exhaust collector gasket. It's made out of soft aluminum, and it cannot blow out. After years of fighting with paper gaskets, this solution is a welcome one. Thanks to Percy's Performance for engineering them.

This is great example of a modern header design. Note the long length of the primary pipes, and the merged slip-on style collector design. When we peek inside the collector, we see how the four primary pipes merge nicely together, and a pointed insert in the center of the four pipes eases the transition from four tubes into one. This pointed insert is called a goilet, and it has become a standard feature in good-quality headers. Also, this header is made from stainless steel, and will not corrode.

Headers have not always been so refined. This vintage pipe shows how far header development has come. The uncoated steel tubing rusts easily, and when we look inside the collector, there's no goilet to be found. Instead, all four primary pipes are crammed together and loosely welded up. This is not conducive to optimum exhaust flow and performance.

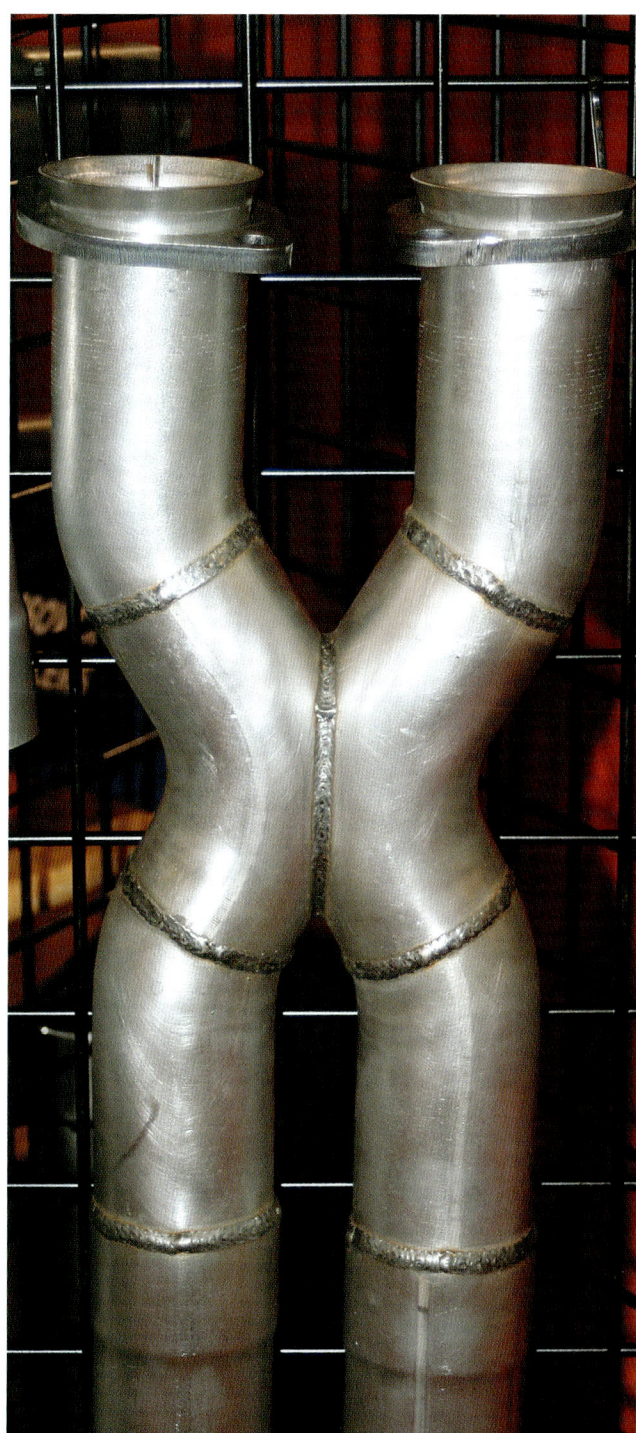

Exhaust research has found power could be had if an X-pipe were installed into the exhaust system just aft of the headers. Now, many different X-pipes are available in a wide range of diameters and configurations.

When we look at the order in which they fire, we see it represented as 1, 8, 4, 3, 6, 5, 7, 2. If we check which side of the engine these firings occur, we see that cylinders 8 and 4 fire consecutively from the even-numbered side of the engine, and later in the firing order, cylinders 5 and 7 also follow each other from the odd-numbered side. If we envision how these exhaust pulses are traveling through the exhaust system, we see the gap in the flow when these consecutive cylinders are firing. Consequently, there's a lack of consistent flow, and the addition of an X-pipe allows the flow from one side of the engine to help keep the exhaust flow stream more consistent from side-to-side.

This smoothing of the exhaust pulses helps pull the spent gases from the engine, helping it work more efficiently. There's power in that efficiency, and the addition of an X-pipe can result in measurable power gains. They aren't huge power gains, but they do exist when compared to an exhaust system with completely independent left and right plumbing. Additionally, an H-pipe simply connecting the head pipes just after the header collectors also helps, but this serves to equalize the pressure from left to right rather than aiding flow. It's better than no connection, but not as effective as a true X-pipe.

The only limitations to X-pipes seem to be finding sufficient space to mount them under the car, but even this has gotten easier with the development of oval-shaped pipe. The additional clearance gained by the use of ovate X-pipes is often sufficient for adequate clearance, even under lowered cars.

If you have a good-fitting set of headers, improved-design collectors, and an X-pipe, you've done about all you can to maximize the performance of a real-world exhaust system. Aft of the X-pipe, system routing must still pass through mufflers, and the choice of which muffler to use is still a matter of personal taste.

Most high-performance muffler manufacturers offer a wide range of case dimensions and pipe inlet and outlet diameters and locations. How you want your car to sound and which mufflers will fit best under your particular vehicle are the considerations you must deal with, and researching these points thoroughly will ensure you get the best-possible results. Many muffler manufacturers now offer sound samples of their products on their websites, and if you're looking for a particular sound, it's worth checking these recordings out. Remember that different headers and camshafts will alter the sound your engine makes, as will X-pipes. The sound you hear on the recording may not be absolutely identical to what you'll ultimately hear from under your own car. Still, some mufflers offer a deeper tone or a louder exhaust note than others.

When shopping for mufflers, look for high flow and low restriction. Luckily, the muffler makers know this very well, and the selection of good-flowing street mufflers today is better than it's ever been, so you should be able to find a good pair of mufflers that fit and sound the way you'd like without adversely affecting the performance of your V-8. In some rare cases, it's even been shown that mufflers improve the power some engines make when compared to open headers. While I'd never guarantee that claim for a street machine with a full exhaust system, the point is that modern aftermarket mufflers are much improved over their vintage predecessors, and you'll lose little if any power by running a great modern muffler.

What about tailpipes? Again, this is a personal call, but know that today's mandrel-bent pipes are much more flow friendly. A mandrel is a flexible insert, which is installed inside the pipe before it is bent into shape. Once the mandrel is in place, the pipe can be bent without crimping, and it can maintain its full diameter throughout the bend. When pipes are bent without a mandrel, they kink a little at the tightest portion of the bend, and no longer maintain their full diameter. This affects flow through the pipe and therefore the performance of the engine. While tailpipes aft of the mufflers will cost a bit of power at some rpm levels, routing the exhaust to exit where you'd like it is less of a performance penalty than ever before.

Deciding what diameter pipe to choose for your exhaust system should be based on the power level of your engine. Typical (300–450-cubic-inch) V-8s that spend most of the time below 5,500 rpm need 2.5-inch-diameter pipes to breathe effectively. If you choose to build a performance engine over 450 cubic inches (or so) or plan to spend ample time in the higher rpm ranges, a 3-inch-diameter exhaust would be recommended.

While these parameters are quite loose, know that routing a 3-inch exhaust system is much more challenging in most cases than fitting a 2.5-inch system. This is even more of an issue if the car is lowered, and also if the owner intends to run the exhaust system with full tailpipes over the rear axle. This is important because exhaust pipes get hot, and if any other temperature-sensitive components are close by (like the fuel system or any electronics or wiring) problems will result. While prefabricated high-performance exhaust systems are readily available for most popular American-made cars, there still may be some fitment issues based on the wide range of production line variations and other modifications that may exist under the car. In these cases, an exhaust professional may be required to ensure everything fits as it should.

An exhaust pro will have the tools and experience to fine-tune the fit of your exhaust system. Many exhaust shops specialize in performance systems, and this is surely something to ask about when shopping for help. Additionally, if no aftermarket exhaust systems are commercially available for your particular application, you'll have to have a system designed and built from the ground up. If the shop you choose specializes in custom fabrication and has mandrels available to make the smooth bends a performance system requires, your odds of being satisfied with the final product go way up.

How well do X-pipes work? This custom-crafted X-pipe setup was designed for use in NASCAR, and if you look closely, this one was autographed by Jeff Gordon. Why would Jeff autograph an exhaust part? It was found to add 5 horsepower to his restrictor plate engine package, and he won the Daytona 500 with this under his DuPont Chevy.

Note the use of oval tubing on this X-pipe setup for increased ground clearance.

It helps to get some local recommendations from other hot rodders in the area when shopping for a competent exhaust fabricator. Any installer should guarantee their work, so any potential issues won't cost you to get fixed. This is not unusual, since custom-fabricated systems might have minor issues. This is one case where having a minor issue (like a rattle) is fairly normal, and the installer should have no problem fixing the issues without charge.

Hot rodders typically like their cars a bit louder than what many consider normal, and increasing the flow potential of an exhaust system will usually make it noisier at some rpm levels. Different regions have different limits on how loud a car can be, and different owners have varying tolerance levels for vehicle noise as well.

Today's high-performance mufflers do a better job than ever before of eliminating noise while maintaining high flow levels. Some designs rely on fiberglass packing material, while others use sound wave reflecting technology inside the muffler case. The fewer turns the exhaust flow stream has to make, the better, so look for a good "straight through" flow path when shopping for mufflers. You may be limited to what mufflers will fit under your particular car, but I've found that the larger the muffler case, the quieter the muffler typically is. Shop for a high-flow, low-restriction muffler that delivers the sound you want. Enough muffler choices exist in the aftermarket for you to get what you're after with regard to fit, flow, and sound.

You'll also find that the materials used in modern mufflers and exhaust pipes are superior to those of the past. Coated-steel and stainless-steel designs offer protection against corrosion and rust-out, and while they may be a bit more expensive, their long life justifies the investment. If you take the time to research and design a complete high-performance exhaust system, knowing it will last long-term softens the financial hit up front.

Chapter 8
The Ignition System

The ignition system is tasked with firing the spark plug at the right time, which sounds pretty simple. This really isn't the case however, and with high-performance engines, the job becomes even harder. Why?

It's a matter of pressure. When the engine compresses the air and fuel, the pressure increases in the combustion chamber and it gets harder for a spark to jump across the air gap on the spark plug. In a high-performance engine, these pressures are further increased, so the job of jumping the spark plug gap becomes even more difficult. More energy is required to make this jump, and this is where modern electronics technology and amplified automotive ignition systems come into play.

This is also why spark plug gap is so critical. Plug gap is measured (typically in thousandths of an inch) between the spark plug tip and the inner surface of the extended electrode. On one hand, we want a big, fat spark to expose as much of the air/fuel mixture to the spark as possible. So, running a large plug gap is a performance advantage. On the other hand, the spark has to jump across that gap reliably at all rpm levels, and a smaller gap makes that job easier. The compromise, of course, is somewhere in the middle. We want to run the biggest gap we can that our ignition system is capable of supporting reliably, so having a lot of spark energy available is the key. It's best to run the stock plug gap at first, and then experiment with slightly larger gaps

Using good-quality spark plugs (like this one from Denso) is an easy call on performance engines. This plug contains iridium, which is incredibly strong and resists wear. It's a well-designed plug that works well in performance engines.

to see if there is any performance advantage. If any misfire develops under high-load or high-rpm situations, close the gap until it goes away. This may seem simple, but it's how you find the largest possible gap you can run without adverse effect.

The spark plug is the final link in the ignition system, and a wide range of plugs have been engineered to perform effectively in all types of engines. So which plug is best for your engine? It will require some experimentation, but the best place to start is with the stock spark plug recommended for your engine. These can be considered the baseline. If there are issues, like plugs fouled by carbon buildup or melting their tips, you can begin experimenting with different plugs.

A lot of new spark plug designs have been brought to market in the last few years, and many of them are very expensive. Are they worth it? In my opinion and experience, I have to say probably not. The basic idea behind the interesting new electrode designs is to provide a lot of surface area for the spark to jump to. Should the electrode become dirty over time and use, greater surface area gives the plug a chance at lasting longer since the odds are good some portion of the electrode will be clean enough to support a good spark. Of course, we're just talking about use in stock, factory engines.

Once we make the jump into high-performance engines, things change a bit. As mentioned earlier, the spark must jump across the gap in the high-pressure area of the combustion chamber, and having a good amount of spark exposed to the compressed air and fuel is a good thing. When a complex electrode is in place, this spark may be shrouded, if it's capable of making the jump at all. If we look at racing spark plugs, like those used in the upper echelons of NASCAR racing, we see a more traditional-looking plug with a trimmed back electrode. This exposes the most possible spark, but there must be a good, clean surface to support that spark. So, fresh plugs are a regular part of race engine maintenance.

The heat range of a plug refers to its ability to dissipate heat from the plug tip into the cylinder head. The more surface area the plug has, the more heat it will hold in the tip. A smaller surface area exposed to the chamber means more heat will be transferred to the head. The optimal temperature range for the plug is 930–1,560 degrees Fahrenheit. On modified engines, the increased cylinder pressure will lead to higher temperatures, so a colder plug may be required if the engine has been modified for increased power.

When reading the plugs, there are some distinct traits to look for. The tip of the plug extends into the combustion chamber, and this is the critical point to read. A good

As the engine increases in rpm, the spark must fire earlier to ensure maximum cylinder pressure at TDC. This is called ignition advance, and nondigital ignition systems common in American V-8s rely on weights and springs inside the distributor to use centrifugal force to advance the ignition timing. Here we see these weights and springs both at rest and open. It's easy to understand how altering these weights and springs will change the ignition advance curve. Fine-tuning the curve is an important part of maximizing engine performance, and experimentation on the dyno to see how power production is affected is the best way to accomplish this.

Ignition amplifier boxes have become very popular as enthusiasts learned the benefits of a hotter spark being fired multiple times. The analog box on the left can be used with any points or electronic ignition setup, and offers internal rev limiters to prevent over-revving and the damage that can result. The digital box on the right includes those benefits and more, as it can be used with programmable ignition systems to fine-tune the ignition advance curve without relying on centrifugal weights and springs. These are MSD products.

running plug will be beige in color, and the tip will be clean. If the tip is coated with black carbon, the plug is not getting hot enough to burn the carbon away and clean itself, so a colder heat range plug would be required. If the tip is burnt and appears damaged, it is getting too hot and needs to be replaced by a cooler heat range plug.

Spark plugs are manufactured in many heat ranges, and these are typically reflected in the part number of the plug. Once you've determined what heat range you need, you can determine the part number of the plug you're after. This can be confusing, because with American-made plugs (Champion, Autolite, etc.), the higher the number, the hotter the plug. For Japanese manufacturers (NGK, Denso, etc.), the higher the heat range number, the colder the plug.

You should also be aware that the heat range numbers used by various spark plug manufacturers are not universal. A heat range 7 plug from one manufacturer may not be the same as a heat range 7 plug from another manufacturer, so either stick with the same manufacturer or research the part number crossover from maker to maker. When comparing

identical plugs with different heat ranges, generally the difference from one full heat range to the next colder range is the ability to remove 150–215 degrees Fahrenheit from the combustion chamber.

Many different materials are used in spark plugs today, and I'd recommend working with standard factory-type plug before investing in some high-tech plugs. Once you've determined the best-possible heat range for your engine, then you're free to experiment with different materials. The exotic materials used in some spark plugs make them quite expensive, so it's not wise to experiment until you're confident in your choice. Platinum makes a great spark, but doesn't last very long. Iridium is a great conductor as well, and will last long-term, but is very expensive. If you stick with traditional plugs in the proper heat range with the best-possible gap dimension, your V-8 will run just fine.

What does this mean to hot rodders with high-performance V-8s? It means you should use good-quality spark plugs and keep them clean and properly gapped. I'd recommend you experiment with the plug gap to ensure

This cutaway distributor shows the weights and springs responsible for the timing advance curve. The weights and spring tensions can be altered to fine-tune the advance, but because they are based on centrifugal force, the curve is linear and must advance timing with rpm. This is an MSD distributor.

This is a distributor designed for use with a crank trigger and a programmable ignition system. You'll note there are no weights or springs at all inside of it, as its job is now to simply distribute the spark energy to the individual spark plug wires, and that's about it! Fewer components typically mean fewer problems, and crank-triggered setups are known to be more accurate too.

In addition to choosing the proper amplifier and distributor, you also have to determine the best possible ignition coil for your application. These two MSD coils have different levels of power and different price points. Getting the right coil to match the rest of your system is as easy as asking for a recommendation from the manufacturer of the other components you're running.

you're running as wide a gap as you can get away with without negatively impacting the engine's performance at all rpm levels. This means you'll have to check the plugs regularly and keep them clean. Such is the life of a high-performance engine, and this should be considered part of the joy of owning one.

Traditionally, the ignition system was triggered by points inside a distributor. As these points opened, the current flowing through them would search for the path of least resistance, which would be through the distributor cap and down the plug wires to the plug. Points are a mechanical wear component, and have to be checked regularly as they wear out to ensure the gap (the distance the points were opened) and dwell (defined as the amount of time they were opened) are correct. The advent of electronically triggered ignition systems (activated by a floating magnet, which does not wear since it never actually contacts anything) eliminated the need for points, and represented a huge step

forward in ignition system technology. Not only did they not wear or require regular maintenance, but they were no longer responsible for transferring all the spark energy that ultimately found its way to the plugs. Rather, they merely triggered the firing of the plug, allowing much greater amounts of spark energy to be used.

This technology appeared in factory electronic ignition systems like GM's popular High Energy Ignition (HEI) system. Use of HEI allowed for larger plug gaps and more reliable spark delivery at high rpm levels. Aftermarket ignition system development took this further by firing multiple sparks instead of one single strike, and even more efficiency was gained. Then, by developing digital controls, an ignition system's advance curve could be tailored in a nonlinear fashion, versus the traditional mechanical controls. In the past, a series of weights and springs would control the advance curve, and the spring-limited weights would rely on centrifugal force to alter the amount of ignition advance.

Good-quality spark plug wires should be part of any performance engine plan, and an investment in good wires assures reliable ignition system operation long-term. These are MSD's 8.5mm wires, which work great with the rest of their quality ignition products.

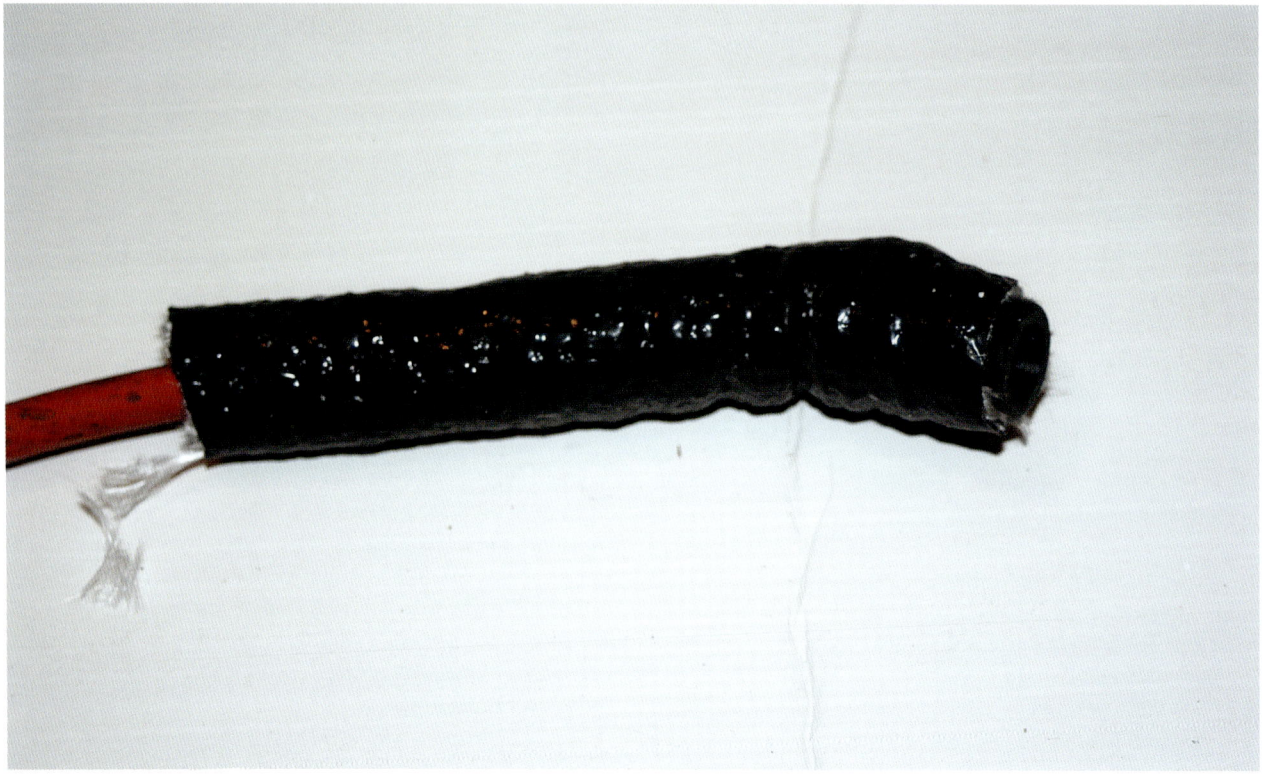

If your plug wires come close to the headers or spark plug heat shields, get some of these boot insulators. They can prevent a plug wire boot from melting and the subsequent grounding out that's sure to follow. These are from MSD and are designed to fit snugly over their plug wire boots.

Have you ever experienced a spark plug wire coming loose at the distributor cap? Most of us have, but adding a wire cap like this one from MSD keeps all the wires on the cap secure. It's one of the little things that can make a real difference.

ignition system ever again, even in restorations where keeping the look of a vintage distributor is desired. Many electronic ignition conversion kits are available to upgrade the internals of vintage-type distributors to electronic triggering without altering their external appearance. This is a wise move, and one I've recommended countless times without regret.

In many street engines, especially those built on a budget, a factory-type electronic ignition can prove perfectly adequate. Luckily, many vintage engines were produced far enough into modern times to be offered with electronic ignitions, and these units are easily adapted to the earlier engines. Even the electronic conversion kits are not prohibitively expensive, so upgrading to electronic triggering is an attainable goal. Furthermore, aftermarket add-ons, like amplifier boxes, can be used with all types of distributors (both factory points and electronic models, along with aftermarket units) and additional spark energy is there for the taking.

Aftermarket distributors are available for most makes and models of traditional American V-8s. Companies like MSD, Mallory, Accel, Pertronix, and Crane have developed excellent distributors whose designs have evolved through racing to a high state of reliability and performance. As an enthusiast, I prefer to choose one company's products and stick with them from the amplifier box to the distributor and even the coil, since these components are designed to be used together as a complete system. I have seen other enthusiasts tie ignition components from different manufacturers together successfully, but by purchasing components designed to work in concert, potential compatibility issues can be avoided. Additionally, if any problems do arise, the single manufacturer you've chosen will be better equipped to assist you with technical questions if all of your components came from them.

In racing, it's common for the ignition to be triggered by a crankshaft sensor. This represents the most accurate way of triggering the ignition, because the timing of the spark is directly related to the crankshaft position. Consider that a distributor-triggered spark is dealing with the effects of the timing chain, the camshaft, the cam gear, and the distributor itself in trying to determine the proper moment to trigger the spark. Each of these components adds a small variable to the triggering event. By referring directly to the crankshaft, all of these variables are eliminated, and the signal is more accurate as a result. If you're looking for ultimate ignition control, investigating a crank trigger is a wise move.

I had the opportunity to see this firsthand in back-to-back testing. I was working in a race engine shop where the primary focus was Ford four-cylinder engines used in open-wheel racing. These engines typically ran 10,000 rpm, so accurate ignition timing was critical. The engine dyno had been wired to run both distributor-triggered and crank-triggered ignitions, and we could switch back and forth while

As engine rpm increased, the weights moved outward at a constant rate and advanced the engine's timing. With a digital setup, one can program the advance curve without relying upon mechanical means, which means you can give the engine proper timing advance at all rpm levels even if it's not perfectly linear.

Choosing your ignition system becomes a matter of determining your own engine's needs and selecting the best possible components, based on your budget. While digital systems command a strong investment, the really good electronic setups are much more affordable than they've been in the past. Even a good factory electronic ignition system is capable of outperforming a points-type system, and points systems offer no performance advantage over their electronic descendents. I can think of no reason to run a points-type

In racing, Hemis were often treated to a second set of spark plugs, as shown here on Dandy Dick Landy's Pro Stock Challenger from the early 1970s. The benefits of dual plugs were a broader spark front and the ability to stagger the timing on both plugs so the second one fires just after the first. The development of multiple-spark firing systems eclipsed the need for dual plugs in all but the wildest nitro applications.

the engine was running. With the timing light on the balancer, there were clearly a few degrees of timing variance when the distributor-triggered system was in use. A flick of the switch over to the crank trigger, and the timing light would seem locked on the proper timing setting, even at 10,000 rpm and beyond. The difference was obvious and impressive, and the benefit of the accuracy of a crank-triggered ignition was a lesson I'll never forget.

When it comes to spark plug wires, I simply invest in the best possible wires available. A good quality 8.5 mm or larger wire is fine for most street and track applications, and issues with plug wires are more commonly related to improper installation or routing than any failure within the wire itself. Buy good wires, assemble them correctly (if they are build-it-yourself types), and route them clear of any heat source—especially the exhaust headers.

If the wires will be close to the exhaust headers, high-temperature insulation coverings are available from many sources, and they do work well. While no part of the wire or insulation should actually touch the header pipe, the addition of insulation can help keep the heat from melting the wire's insulation or the spark plug boot.

Chapter 9
The Cooling System

Keeping your V-8 at the proper operating temperature is essential, and once you've begun modifying it, this can become a challenge. Luckily, modern advances in technology and equipment have made this easier than ever before. If a well-built V-8 has properly functioning cooling system components, it should be capable of maintaining a regular operating temperature. If it is not, either some components are not functioning correctly or there are inadequacies in the capability of those components.

Once you modify an engine for increased performance, the requirements for cooling it can change as well. If you increase the compression ratio, the engine will generate more heat, which must be dissipated by the cooling system. If you

regularly run the engine at or near its rpm ceiling (redline), it will inevitably generate more heat. If you operate the engine regularly at WOT, it will generate more heat than it was originally designed to.

With increased heat generation comes a need for improved cooling capability. This means you'll have to move more coolant through the engine and radiator (through use of a higher volume or higher efficiency water pump) or improve the capability of the existing radiator (by upgrading to a radiator capable of shedding more heat, like an aluminum unit, or a greater-capacity radiator containing more cores). You could also increase the size or efficiency level of the cooling fan(s) by upgrading them; this will have its greatest

There are good quality aluminum radiators made to fit just about every domestic vehicle ever made, and some manufacturers will even make one to your specifications if one is not available. This unit replaced the factory brass unit perfectly, and hooked up without issue. Investing in a good radiator like this will help protect your

Here's a good example of an aftermarket aluminum radiator outfitted with a matching electric fan. The electric fan has an integral shroud to help its efficiency, and when compared to the in-car installation shown on previous page, is a much cleaner package. The fan shroud, while better than nothing, does not fit closely to the fan blades. Conversely, this electric fan fits tightly to both the radiator and the integral shroud, and it will be more efficient in moving air because of it.

effect at low vehicle speeds. Once you've achieved freeway speeds, the amount of air passing through the radiator should be sufficient to cool the engine. If the engine cannot cool adequately at freeway speeds, the fan(s) are not the problem.

Fans can be either mechanical (engine driven) or electric. They vary widely in design and construction, and my recommendation here is to base your decision on the specific need. Performance engines typically generate more heat than their production line counterparts, and sometimes they require core cooling because of it. If your factory fan design proves to be capable of keeping your engine cool, there may be no need to change it. While some fan manufacturers claim

power gains from the addition of their product, these gains are typically trivial. The cooling system's job is to maintain engine temperature within a predetermined range, and unless it cannot accomplish this task, I see no reason to upgrade.

If you find the need to upgrade, aftermarket fans with more fan blades or a more extreme angle on the blades will draw more air through the radiator. The fan's efficiency can be increased through use of a good-fitting shroud, which should fit snugly to the radiator. Optimally, the fan's blades should, while spinning, extend partially into the shroud. A larger-than-stock fan diameter may pull more air, but will require a correspondingly larger shroud.

Another good investment is a top-quality high-flow water pump. Many are available in the aftermarket for a wide range of domestic V-8s. Simply put, they are capable of moving more coolant than the factory units they replace, and are typically produced to a higher quality standard as well. The way I see it, your high-performance V-8 needs additional cooling capability, so it pays to get a good water pump now rather than when the stocker fails you and leaves you stranded or damages your engine.

Many enthusiasts are replacing their factory mechanical fan setups with electric units. I prefer the electric fans that come with well-engineered shroud assemblies as part of an integral package. Some manufacturers like Flex-A-Lite offer complete high-performance cooling system solutions (radiator, fan, and shroud as a matched package) for more popular applications.

Electric cooling fans can provide a remedy for overheating, since they can be programmed to turn off and on at predetermined temperatures with a thermostatic controller. This is convenient, because if overheating problems persist, you can adjust the thermostat to come on earlier and stay on longer to combat the heat. For example, if you wanted to maintain 190 degrees, you could adjust the electric fan to come on at 190 degrees and turn off at 175 degrees. If this didn't maintain your target temperature, you could adjust the fan to come on at 185 degrees and turn off at 170 degrees. Fine-tuning in this way offers a solution for just about every cooling system.

The addition of air conditioning systems (A/C) is popular in street machines today, and while it makes for a comfortable interior, it adds even more load to the engine. If you choose to upgrade to air conditioning, you should be prepared to upgrade your cooling system to compensate for the extra load.

A wide range of radiators are available too, and a custom radiator can be built for virtually any application. Aluminum radiators are very popular as factory replacements, and they can be had in with additional cores (when compared to the stock parts they replace) to increase coolant carrying capacity. Brass radiators are still made and do a fine job of transferring heat, but the new aluminum units weigh less and seem to have a wider range of options nowadays.

A well-designed performance car should be able to idle comfortably without overheating. While it may get a little warmer sitting in traffic on really hot summer days, it still should never reach a point where it actually boils over. A 50/50 mix of water and coolant won't boil until it reaches more than 230 degrees. In hotter climates, this ratio can stand a bit more water than coolant to gain some effectiveness, since water transfers heat more effectively than coolant. I've run as much as 80 percent water in some applications during the summer in Southern California without issue.

If cooling is still a challenge, there are some coolant additives available in the aftermarket (Red Line's Water Wetter and Royal Purple's Purple Ice products are two examples) that can help. They may only push the boiling point up a few more degrees, but that might be all you need.

A typical thermostat opens at 160–180 degrees, leaving plenty of room for the engine to heat up a bit. The key is to have the engine maintain 180–190 degrees during normal operation, and the cooling system should be capable of this.

If the water pump is incapable of moving enough coolant through the cooling system, the engine will gradually get hotter until it overheats. Higher-efficiency water pumps employ tighter clearances, larger fins, or larger passages to move more water than their stock counterparts, and can really help overheating issues.

The thermostat is a simple temperature-controlled valve, one that is essential if a street car is to maintain a steady operating temperature. Some enthusiasts have claimed to remove the thermostat entirely and never have a cooling issue. While this is possible, it's also quite suspicious. It's easier for an engine to get to operating temperature and maintain that temperature with the correct thermostat in place. I typically recommend a lower-temperature thermostat only if a car has cooling issues, as it should allow for cooler operation if all else is right.

Modern late-model V-8s are designed to run at even higher temperatures (205–210 degrees) to aid emissions, and these engines work just fine at these temperatures. While some enthusiasts have added higher- or lower- temperature thermostats in a quest for greater performance, I've not seen consistent, reliable results to justify such a modification. My experience has been to stick with factory recommendations on thermostat temperatures, and the engine should be capable of heating up and cooling down without issue. If there are problems, by all means they should be diagnosed and solved, but I've never found a reason to recommend a hotter or cooler thermostat for improved engine performance.

Competition cars typically run no thermostat, but will use a restricter plate in its place. Because racing cars remain in motion almost constantly, they have no need for a thermostat. Since street cars can spend as much time sitting still at idle as they do at freeway speeds, a thermostat is a requirement.

Chapter 10
The Fuel System

The purpose of the fuel system is simply to store and deliver fuel to the engine as it is required. This seems simple enough, but the fuel systems designed and installed in most vintage American V-8-powered cars and trucks are typically not capable of supporting a modern high-performance engine. This is especially true when on-track situations arise, and the vehicle sees extreme forces it was never designed to deal with. By isolating each component in the fuel system and reviewing it, we can develop a sense of what is truly required to support a high-performance engine.

Starting with the fuel tank, we'll need to ensure fuel will always be directed to the fuel pickup. This typically means some type of baffling should be installed to guide the fuel and prevent g-forces from pulling it away from the critical pickup point. When looking at your fuel tank, is the fuel pickup at the absolute lowest point in the tank? It should be, as gravity will naturally want to pull the fuel there, but you may be surprised to see this is not always the case. Secondly, are the fuel lines of sufficient diameter to feed a hungry V-8? Most factory tanks rely on a 5/16-inch-diameter fuel line, which may be capable of feeding a 350-horsepower engine at WOT. Many of today's performance V-8s make much more than 350 horses, and an upgrade to 3/8-inch-diameter or even 1/2-inch-diameter fuel lines is necessary to ensure

This is an aftermarket stainless-steel fuel tank from Rick's Hot Rod Shop (El Paso, Texas, 915-760-4388, rickstanks.com). In addition to never corroding, these tanks can be equipped with internal electric fuel pumps ready to work with an EFI system. Rick's also offers some tanks with additional fuel capacity, and others narrowed to fit between widened wheel tubs.

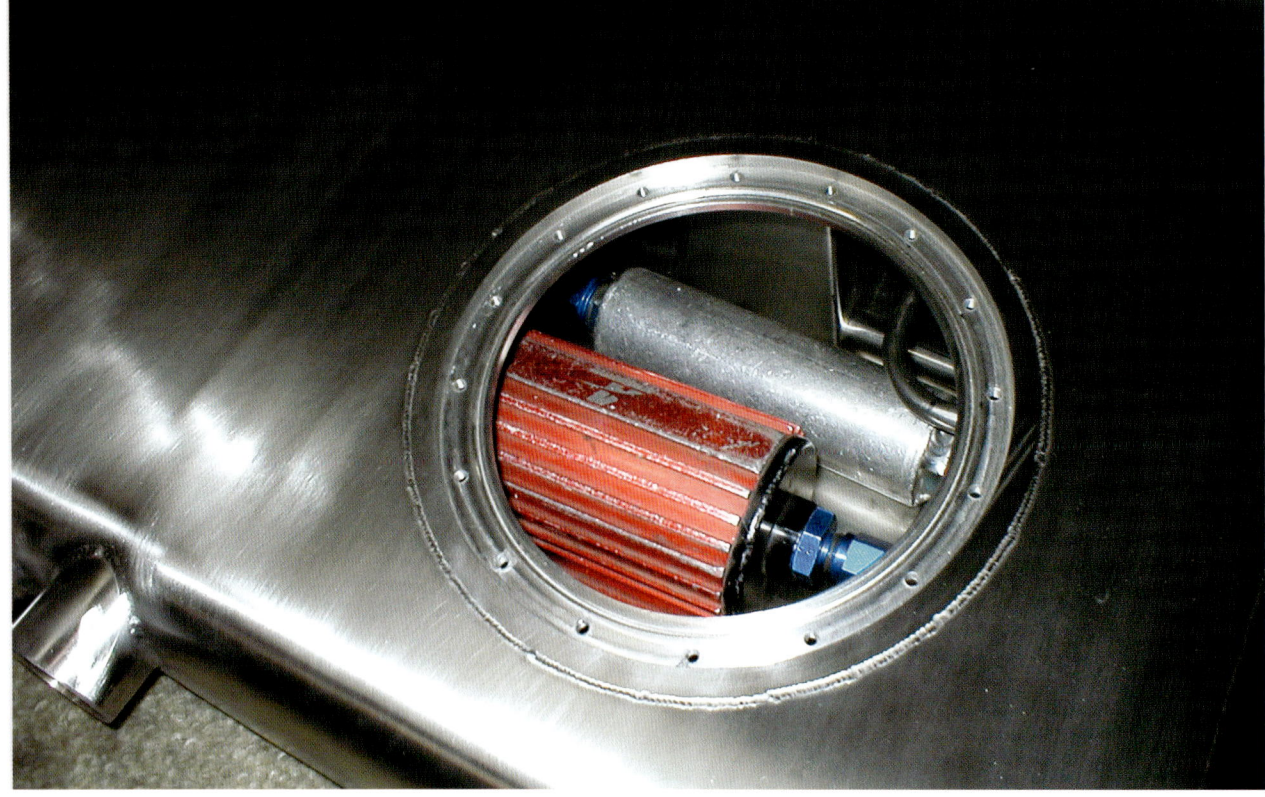

A peek inside the access port of a Rick's tank shows the electric fuel pump and the fuel filter assembly. The GM fuel pump will support 450 horsepower. Naturally, fuel pump upgrades are available for higher horsepower applications. The in-tank design is very quiet, and the GM pump is proven reliable.

adequate fuel volume and pressure at WOT. This is especially true if you plan to be at WOT for extended periods, like on a road course or at an open road race.

Luckily, the solution is relatively simple for most situations. While aftermarket fuel cells have been available for a long time, they don't typically fit directly into place where the factory tank was, and often a smaller-capacity fuel cell becomes the compromise solution. If a fuel cell is not required by rule for your project vehicle, I'd encourage you to investigate other options.

Reproduction fuel tanks are offered for most popular V-8-powered vintage vehicles, and a wide range of top-quality stainless-steel factory replacement tanks are also finding a market. The benefits of buying a new or replacement tank include the ability to add baffling, modify or replace the factory fuel pickup sump, and the chance to upgrade the fuel pickup to a larger-diameter line. Many enthusiasts are also opting for return lines back to the tank to enable pressure-regulated carburetor systems or EFI setups.

Setting up a new fuel tank with internal baffling, a proper sump, and adequately sized supply and return lines can all be accomplished simultaneously with the purchase of a new tank. Naturally, your old tank could be removed, cleaned, modified, and resealed to accomplish the same task. Personally, I prefer to start with new tanks because I know they can be welded safely,

and there's no chance of discovering any leaks or corrosion. Considering the affordability of most reproduction tanks on the market today, modifying the original tank doesn't make financial sense either. This can only be justified if a reproduction fuel tank is not available for your vehicle of choice, and the one you've got is in suitable condition to be repaired and/or modified for true high-performance use.

Drag racers should be focused on the rearward thrust caused by the sudden launch at the starting line. Baffling should be designed to keep fuel from moving rearward, and the sump should be designed to deliver fuel aft from the lowest point in the tank. Street and handling enthusiasts should be concerned with fuel splashing from side to side in turns, and their baffling should be designed to prevent this. The sump, again at the lowest part of the car, can face either fore or aft, but I prefer a forward-facing sump with the lines plumbed into the front portion of the tank here. It should be a cleaner setup, and the fuel lines won't have to perform a 180-degree turn to get the fuel moving back toward the engine again. In a drag car this is simply part of the program, and should be taken into consideration at the design phase.

In developing a performance fuel delivery system, line routing and placement become critical. Making 180-degree turns should be accomplished as gently as possible, because liquid fuel flows more smoothly around gentle radius turns

This unfinished tank allows us to see the internal baffling being installed at Rick's. By building dams like this inside the tank, the fuel is not allowed to slosh to one side while cornering, or to the rear of the tank under hard acceleration. This ensures fuel is always available to the pump pickup.

High-performance electrical fuel pumps have been available to enthusiasts for a long time, but now some great mechanical pumps are offered as well. This one is from Stewart Components, long known as a maker of race-level water pumps and other things. This pump provides ample pressure and volume to feed the hungriest of engines.

A good fuel pressure regulator is necessary if you're running a high-volume, high-pressure pump, or if you've got a wet nitrous setup pulling fuel from your system. Regulators are standard fare in most EFI systems too. The regulator should be located as close to the carb or fuel rail as is feasible, so the pressure setting is as close to what's going into the carb or fuel rail as can be.

than tighter ones. Too tight of a turn can reduce fuel flow and impact pressure, which is precisely what we're trying to avoid. Gentle turns are best, even if it means having to run a longer length of fuel line. The distance the fuel must travel is less critical than its path.

The next consideration is the fuel delivery system components. These include filters, pumps, and in some cases, coolers. All new cars and trucks use electric fuel pumps, and these have proven to be durable and effective at providing adequate fuel pressure and volume. By locating them inside the fuel tanks, the car manufacturers have managed to make them quiet too. Such engineering is worthy of consideration by enthusiasts as well. If a factory pump can suit the needs of your project vehicle, and can be mounted inside your new tank, this setup can be as reliable and effective it is in a new car. You'll be spared the task of locating and mounting an external electric fuel pump, and you won't have to listen to it run all the time.

The only downside to this is that you'll have to remove the fuel tank to repair or replace the pump should a problem arise. But, if the benefits outweigh this potential problem, this is certainly a good option. The wide range of in-tank factory pumps available means one of them should be able to suit your needs. Even if a given pump was designed to feed a fuel-injected engine, it could be possible to regulate

the pressure down to the low levels carburetors need. Some research on your specific application is necessary, but it could pay off with a simple, effective, and quiet fuel delivery system.

If you choose to go with an external electric pump, plumbing a fuel filter between the tank and the pump inlet is a requirement. I suggest getting a good-quality filter with either a replaceable filter element (like a small oil filter) or one with an easy-to-clean filtration medium. The fuel filter should be plumbed with the same diameter line as the fuel feed line so no flow or pressure is lost through the filter, and the same is true for the fuel pump. The fuel system will be limited by the amount of fuel that can flow through the smallest-diameter orifice it contains, so it makes little sense to choose a fuel filter (or pump) with 5/16-inch-diameter fittings if you are upgrading to 3/8-inch- or 1/2-inch-diameter line.

There are many good-quality electric fuel pumps on the market today, and there's little reason not to be completely satisfied with your purchase. Most fuel pump manufacturers rate their pumps based on how much horsepower they can support, and this is a big help to enthusiasts. Fuel pumps are traditionally rated by how many gallons of fuel they can move in an hour (gallons per hour, or gph), but few engine builders know how much fuel a powerplant will need over

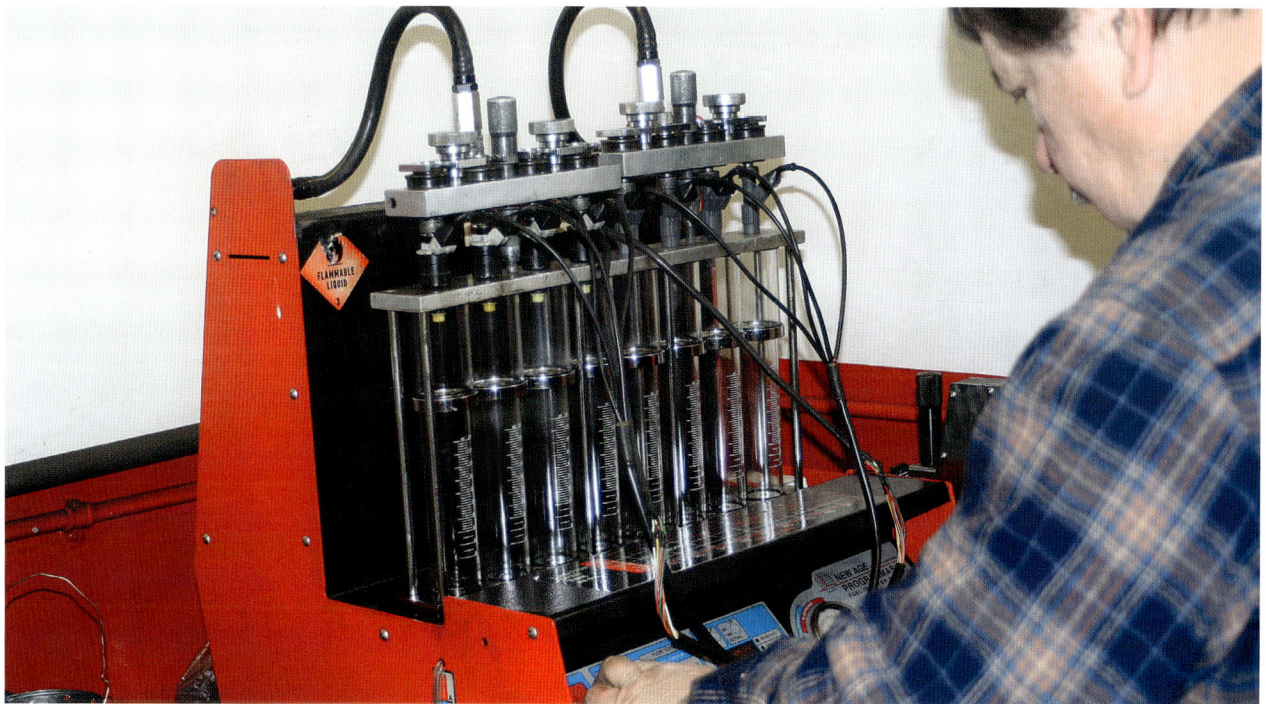

This machine tests the function of fuel injectors and measures their flow rate in timed cycles. The results verify the flow ratings of the injectors, and ensure each one functions as it should.

an hour of run time. Naturally, gph ratings are also affected by the diameter of the lines used in the testing, and odds are good that the largest-possible lines were used. By using horsepower ratings in addition to gallons gph it's easier for enthusiasts to find the best possible pump for an application.

You should also know that the voltage going to the fuel pump will have a direct impact on its performance. The pump should see a clean 12-volt signal at all times, and fuel pump relays are common for this reason. I use and recommend fuel pump relays for all aftermarket electric fuel pump upgrades.

Many enthusiasts still rely on mechanical fuel pumps, and these have plenty of merit. Engine-driven mechanical pumps have enjoyed modern evolution and technology as well, and many are opting to have an electric pump at the rear (near the fuel tank) feeding into a mechanical pump (on the engine) to ensure fuel is being fed steadily to the carb. This is a good idea for many street enthusiasts who prefer the look or reliability of the mechanical pumps.

Still others feel a good-quality electric pump is all that should be required, and having a mechanical pump inline is just one more thing that could go wrong. I feel this is a personal preference, and as long as the engine is fed a steady diet of fuel in all driving situations, it will be fine. I don't choose to run mechanical pumps alone anymore in any truly high-performance situations, but I've known guys who do without issue. This is especially true at the drag strip, where street cars have gradually morphed into race cars and the fuel system has

yet to fail, so it's yet to have been upgraded. There is a limit to what stock-type mechanical fuel pumps can do, and I'd rather not find the limit at the top end of the track. At the very least, upgrade to a high-performance mechanical pump. For more popular engines, these are common, but for oddball makes, the step up to an electric pump might be the only option.

Once the fuel reaches the engine, it should be regulated to ensure consistent pressure and volume. While fuel pressure regulators have been a part of hot rodding for decades, the latest versions are more consistent and reliable than their predecessors. The regulator should be mounted as close as possible to the carburetor's fuel inlet to limit the chance of anything changing between the time the monitored flow comes out of the regulator and heads into the carburetor.

EFI systems, fuel pressure levels are greatly increased when compared to carbureted setups. Where a carb typically requires 6 to 8 psi, EFI fuel system pressure requirements range from 40 to 60 psi. With EFI, electric fuel pumps and regulators (with return lines back to the fuel tank for bypassed fuel) become the norm. Naturally, the requirements for pumps, fuel lines, and regulators all must conform to the need for higher pressure.

I mentioned fuel coolers earlier, and these are becoming more popular of late. The reason is that in many high-performance fuel systems with return lines, fuel is constantly being circulated from the tank to the regulator and back again. Systems with return lines are much more common in cars with EFI that rely on high fuel pressure to spray fuel into

The four injectors on the left are being compared to the four on the right. This test also allows their spray pattern to be observed. Before installing fuel injectors on your new engine, it's wise to have them flowed and tested in this manner.

the engine. With every trip from the tank to the regulator and back, it picks up a bit of heat from several sources, including the pump and the fuel lines themselves. This is especially true as the lines get closer to the engine and its radiant heat. Over time, as greater quantities of fuel make the trip through the recirculating system, the entire fuel load gets warmer and warmer, and can actually cause vapor lock if the fuel boils in the lines. This can become a major problem on long trips or during a long day of racing. The addition of a fuel cooler to shed this acquired heat is a wise move on any constant-flow fuel system with regulated fuel returning to the tank. After hearing repeated nightmares about this, I must recommend a fuel cooler for anyone using one of these systems. A simple inline finned heat sink cooler is all that is required in most cases to solve the problem.

FUEL INJECTOR SIZING

When upgrading to EFI, it's critical to determine the size of the fuel injectors you'll need. There's a formula to figure this out, and it's not too complex.

Injector sizing is based on the needs of the engine, so we must have a target horsepower number, and an estimate for how efficiently the engine uses fuel. We'd need a term describing how much fuel you use per horsepower, per hour. This is described as brake-specific fuel consumption, or BSFC.

We also must understand how the fuel injector works. Basically, it is a normally closed electronically controlled valve, which (when energized) opens and allows fuel to pass through it. As engine rpm increases, so does the number of times the injector must open and close. Injectors come in different sizes, and larger injectors don't need to be open as long as smaller injectors when feeding the same engine. As a rule, fuel injectors should not remain open more than 80 percent of the time (called duty cycle), and this is why proper injector sizing is important. Too small of an injector may be asked to work more than 80 percent of its duty cycle on a large engine making big power. Too large of an injector will sacrifice efficient performance at low rpm, when it is challenged to slow down its opening and closing rates to minimize fuel delivery. A properly sized injector will be capable of working effectively throughout the rpm range, providing good idle and WOT performance.

To determine the proper injector size, you need to know the target horsepower, how many injectors you'll be using (one per cylinder, so eight for a V-8), your BSFC (0.50 is the standard for naturally aspirated engines, while supercharged and turbocharged engines typically use a 0.60 factor), and the peak duty cycle number (80).

The formula is: horsepower x BSFC/number of injectors x duty cycle = injector size in pounds/hour.

If we are building a naturally aspirated V-8 with a target horsepower of 400, the formula looks like this:

$400 \times 0.50/8 \times 0.80 = 200/6.4 = 31.25$ pounds/hour

So 31 pound/hour injectors would be required in this application.

Chapter 11
The Performance Package

DETERMINING YOUR NEEDS

Throughout this book, you've read about the need to develop a comprehensive performance package, as it really is the key to making big power. Matching critical components from top to bottom is what it's all about, and the more specific you are about your goals, the closer you'll come to reaching them.

Many enthusiasts feel that using race-level components means they'll have race-level performance on the street, but this is simply not the case when it comes to internal engine goodies. Consider that a race car is designed to run very hard for a short period of time, and that it spends most of that time at high rpm. On the other hand, a street car is designed to run for a long period of time, and most of that time is spent at low rpm. For this reason, race-level cams, carburetors, intakes, cylinder heads, and big-tube headers are typically of little use on the street. In fact, developing a well-mannered street car is more difficult than designing a pure racing engine. Getting a suitable idle, tuning for off-idle and part-throttle performance, and keeping it all from overheating is not an easy task!

A quality engine dyno with a skilled operator is a powerful tool. By testing the engine in this manner, we are not only able to determine its power levels, but also fine-tune the engine to ensure it's working to its fullest potential. It's also a great place to test any power parts we might be considering. This Superflow engine dyno at TPI Specialties is fully equipped to test just about any engine, and the years of experience the crew has working on it pay off for their clients.

Tuning for Maximum Performance

The value of engine dyno testing cannot be overstated. Not only are you able to get firm, repeatable data on your engine, but you'll have the chance to tune it to perfection long before it sees duty under your hood. Any leaks can be fixed while you have good access to them, and any problems can be diagnosed and corrected before the engine gets installed. Any engine you've taken the time to design and build is worth dyno testing, and anyone who claims engine dyno testing is worthless or a waste of time and money doesn't know what they're talking about. The advantage of knowing your engine is properly tuned, jetted, and timed is crucial. Even knowing that it starts and runs is important.

Let's say you choose not to dyno test your new engine. You install it in the car, turn the key, and nothing happens. Now, is it a problem with the car or a problem with the engine? Had the engine been dyno tested beforehand, you'd know it ran well. If it didn't start once installed in the car, you'd know the problem must be in the car. If this same nontested engine was installed in the car and started up nicely, but had a leak or a knock or some other problem, this would be easier to correct with the engine out of the car. Personally, I don't like to install an engine until I know it's as close to perfect as possible.

Rear-wheel dyno testing is another big step, and I also feel this is well worth the investment. Once a dyno-tested engine is installed, the tuning requirements may change a bit from what they were on the engine dyno. There are probably different headers installed, and the way a transmission loads the engine is different from the way an engine dyno loads it. Some minor fuel or ignition timing tweaks may be required, and this should be done on the rear-wheel chassis dyno. Once a vehicle has been fine-tuned on a chassis dyno, it can be driven confidently to its full potential, knowing the tuning is as precise as possible.

Both rear-wheel chassis dynos and engine dynamometers use exhaust-based monitoring to determine optimum tuning and performance. The amount of fuel going into the engine is compared with the oxygen content of the exhaust coming out of the engine to determine how efficiently fuel is being used. Variables such as altitude can make a big difference in how much fuel your engine can use. At higher altitudes, there is less oxygen in the air, so less fuel is required to maintain a correct air/fuel ratio. Even humidity can make a difference, which is why it's common to see drag racers with weather stations in the pits. Fine-tuning the jetting to achieve maximum performance requires some experience and the proper instruments to measure how much oxygen is in the air and how much oxygen is in the exhaust pipe when the engine is running.

Having these tools and knowing how to use them is becoming more common, but if you don't have them, it's better to run a little rich than a little lean. A lean-running engine will get hot, and could potentially cause damage. Running a little rich will waste some fuel, but the engine will run cooler and won't hurt itself. Most old-school hot rodders still read their spark plugs to determine whether their engine is running rich or lean, and in most cases this method is perfectly acceptable. A black or wet plug is too rich, where a white plug is running too lean. A tan color on the plug is a good sign the air/fuel mix is right on.

When shopping for a dyno shop, whether it's an engine dyno or a chassis dyno, it's important to look beyond price. All dynos (and dyno shops) are not created equal, and you want to work with the best possible technicians with the best possible equipment.

It's best to stop by the shop you're considering in person to ask a few questions and have a look around. Your dyno shop of choice should be clean, the equipment should be new (or very nicely maintained), and there should be a fair amount of traffic coming through the shop. Ask if you can visit while another dyno test is being accomplished, and observe the procedures being done. Note how safely the technicians work, and determine if their standards meet your own. You should be impressed by the efficiency and thoroughness of the process, and the data gathered should be repeatable and consistent (if the engine or vehicle being tested is of high quality, which also says a lot about the dyno shop in question).

Prices for dyno testing average around $100 per hour, and you should expect to spend at least three hours per session. Naturally, any tuning or repairs will extend this time, and the clock is running the whole time you're there. This is why it's important to observe how the shop works before you choose to spend your hard-earned dollars there. If the crew works effectively and efficiently, you know you're getting the most for your money. If you observe high-level performance and racing vehicles being effectively tuned by a group of enthusiastic technicians, odds are good you're in the right place. If a single technician is working in a dirty dyno on a leaky engine, you may want to look elsewhere.

Maintaining a Performance Engine Long Term

High-performance engines require more care than stock engines, because they are expected to work harder and generally have more adjustments that need to be checked regularly. Things like valve adjustments (on solid lifters, both flat-tappet and roller designs) and regular inspection of both ignition timing and carburetor jetting become part of life with high-performance engines. For those of us who truly love having a high-performance engine running at its peak at all times, such maintenance becomes a labor of love. Owning something as special as a well-designed high-performance V-8 makes these maintenance tasks less of a chore and more of a pleasant way to spend an afternoon.

How often you should perform maintenance tasks is directly relative to how often you use your high-performance car, and at what level. If your car only gets driven on nice weekends in the summer, you should be able to check everything right before the summer driving season and be fine. If the car is driven daily year-round, it should be checked thoroughly every three to four months. Should

TPI Specialties has equipped their dyno with eight individual wide-band oxygen meters. You can see how each of these meters is hooked up to a sensor placed in the header primary pipes. This allows them to monitor and tune each cylinder individually, and is much more precise than relying on one single sensor placed in one header collector. Monitoring each individual cylinder is especially important with EFI engines, since each cylinder has its own fuel injector.

the car be used for competition in addition to street driving (like a weekend drag racer or autocross car), even more frequent maintenance may be required before heading to the track.

It's wise to keep a maintenance log on what you've done, when you did it, and what adjustments or changes have been made. I use a small pad of paper for this, and keep it in the glove box so it doesn't get lost. My carb and timing adjustments, spark plug experiments, and valve adjustments are all recorded here, and when I head to the drag strip, all the relevant performance information is also recorded. This is especially important when fine-tuning at the track, since any changes are made in an effort to improve the car's performance. By recording the change made and what effect it had, I can refer back to it in the future. For this reason, only one change should be made at a time when pursuing peak performance. Maintaining a performance log helps track the effectiveness of any changes, and is highly recommended.

Many enthusiasts don't drive their hot rods during the winter months, and storing the vehicle brings its own maintenance requirements. In addition to keeping the battery charge up and fresh gas in the tank, it helps to turn the engine over occasionally to make sure the same valve springs don't stay compressed for months. It's common for race engine owners with extremely high valve spring pressures to back off the adjustments on their rocker arms

to relieve the pressure on the springs when the engine won't be used for a while. For street cars, this may be overkill, but the idea is the same.

If you're storing the car for the winter, ensure the coolant system has sufficient antifreeze. It's also a good idea to mouse-proof your stored car by plugging the exhaust pipes, closing all the windows, and placing a couple of traps under the car as well. Some enthusiasts prefer to store their cars on jackstands to keep the weight off the tires and suspension. These become the best way for rodents to enter the car, and they should be fortified against critters too. Rodents love to chew on underhood wiring and nest in seat stuffing, so taking steps to prevent them from entering the car is wise. It's also a good idea to check on the car periodically throughout the winter to make sure no infestation has occurred. Rodent droppings are easy to spot (they typically look like dark brown rice) and if you spot them under your car or on its interior carpet, you've got problems. Solve them as soon as possible, since rodent damage is expensive to repair.

When pulling the car out of storage for use, check all the basic fluid levels and air pressure in the tires before heading out for a drive. If you've maintained the car correctly, the first drive should be drama free. Giving the car a thorough detailing not only makes it look better, but also gives you a great chance to look everything over and find any leaks or evidence of rodents you may have missed before.

Chapter 12
High-Tech Treatments and Materials

The process of cryogenically treating engine parts destined for use in high-performance engines has received some recent attention. In this process, the engine parts are heat treated, and then dry frozen to 300–320 degrees below zero Fahrenheit before being assembled. The purpose of the freezing is to force the parts to shrink to their smallest-possible size. This process forces the granular structure of the part to align from the shrinking, and this alignment remains intact once the part returns to room temperature. The crystalline granular alignment helps make the parts stronger and increases their durability. This treatment also relieves stress on the parts that results from the manufacturing processes that create them. The casting, machining, and heat most high-performance engine parts are exposed to can dramatically affect their molecular structure, and cryogenic treatment can create a stronger, denser, more uniform molecular structure.

Reducing the stresses in relatively heavy parts that are subject to high levels of heat and pressure provides the greatest benefits. Performance and durability gains have been seen from the treatment of engine blocks, crankshafts, connecting rods, pistons, cylinder heads, camshafts, valve springs, lifters, pushrods, valves, rocker arms, intake manifolds, and even some driveline parts and brake rotors. It is an added expense, but like insurance, it may be well worth the investment for some.

Similarly, a new vibratory process is being used to treat engine parts prior to severe use. In this process, parts are bolted securely to a metal tabletop, which vibrates to shake the granular alignment into shape. Vibratory stress relief

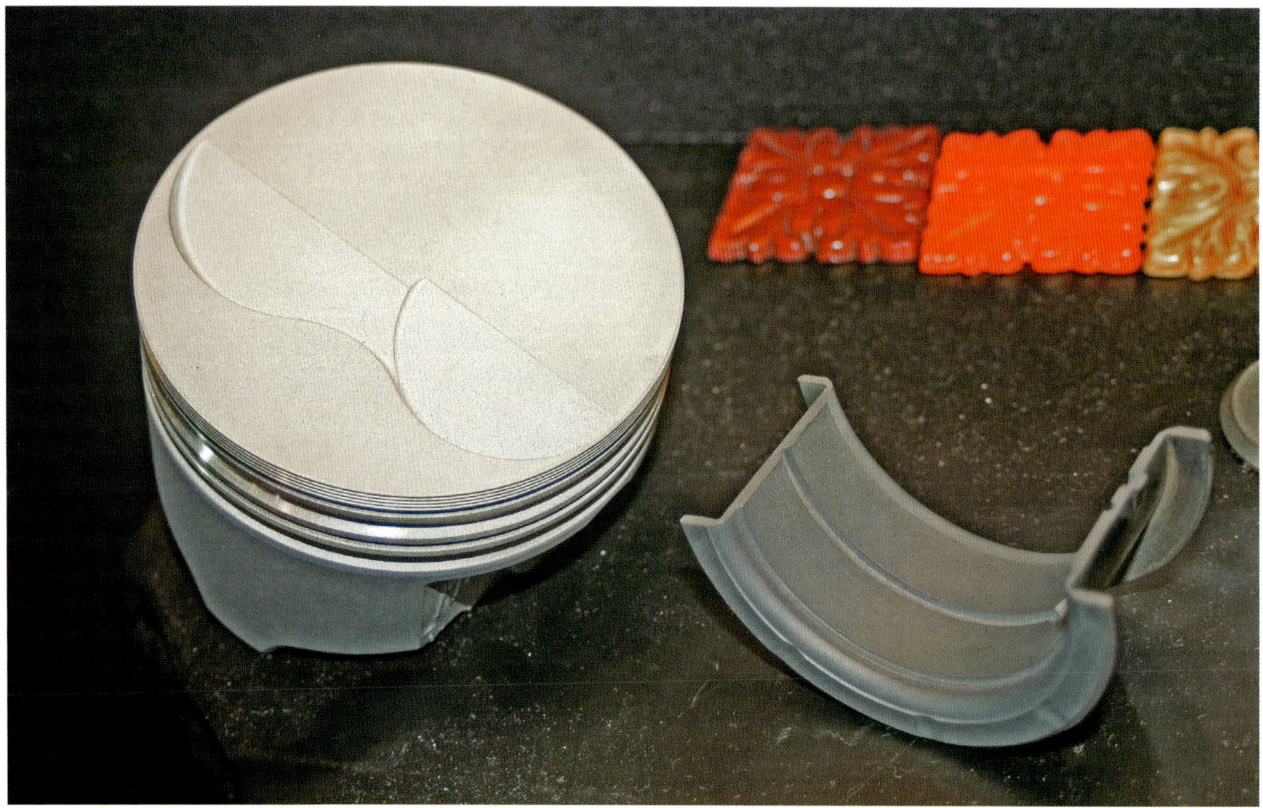

High-performance internal engine coatings have been popular in racing (and the aerospace industry) for many years, and are now finding favor with high-performance street applications. Shown are a coated piston and some coated bearings, which offer both performance and durability benefits.

The graphite coating on this piston reduces wear, friction, and the amount of heat it generates as it moves up and down the cylinder bore. It's possible to purchase many types of pistons with these antifriction coatings applied at the factory, and they really do make a difference. I expect such coatings will become standard fare in the future. These coatings are from Tech Line Coatings (Murrieta, California, 972-775-6130, techlinecoatings.com), and represent only a small sampling of their products. Shown are both their solid film lubricant (which was used to coat the skirt of the piston) and their oil-shedding thermal barrier, which is used to aid in removing oil from moving parts, like the crankshaft counterweights. You can apply these coatings yourself at home or have an experienced pro do them for you.

uses low-frequency, high-amplitude vibrations to reduce a part's residual stress level to a point where it cannot cause distortion or other problems. The frequency the part is treated with and the time it will require for treatment varies with the size and strength of the part.

Both cryogenic and vibratory treatments have shown long-term durability benefits, and if you're making a serious investment into a performance engine, it makes good sense to look at treating at least the engine block and reciprocating assembly. If the parts you're working with are rare or were custom-crafted for your specific engine project, this makes even more sense.

When designing and building a custom high-performance V-8 engine, you invest plenty of time and effort into developing a complete performance package. After your very best efforts, and after significant investment, the power and torque numbers deliver the verdict on the quality of your efforts. After you're done, and your best efforts have been put forward, you should be confident there isn't any more power to be easily found in the design. But, unless you've used the latest high-tech coating products, you may be wrong.

High-performance coating technologies were developed in the aerospace industry to add durability and reduce friction in aircraft and spacecraft. Initially, these technologies carried a high price tag, but as time passed these same technologies became more common, and the prices began to drop. Today, these products are readily available and can be affordable.

There are several different kinds of performance coatings available to accomplish different tasks. Of particular interest to hot rodders are antifriction coatings, thermal barrier coatings, and heat-shedding coatings. When strategically applied to specific engine parts, these coatings can make a measurable difference in power and add durability.

The most common performance coatings are probably of the antifriction variety. When applied to piston skirts, engine bearings, and valve springs, these coatings minimize the friction between metal engine parts dramatically. The energy not wasted on friction now goes to the fly-wheel instead.

It's now possible to purchase precoated engine bearings in all sizes, and testing has shown these coated bearings to not only minimize friction, but also to extend bearing life. Both main bearings and connecting rod bearings are commonly coated. Many piston manufacturers also offer products with coated skirts, and these have shown minimal wear after many miles on the road while minimizing the friction between the skirt and the cylinder wall. One of the more dramatic applications involves valve springs, which are notorious for generating lots of heat as the wire coils contact each other every time the spring is compressed. Repeated heat cycles also weaken the valve spring, and the coated springs have shown extended life when compared to noncoated pieces. The reduction in heat generated by coated springs could justify the cost of the coating by itself, but the extended life given to these same springs should make the choice to coat them simple. With the popularity of roller cams on the rise, and the additional spring pressure these cams require, I'm confident coated springs will soon become a common requirement rather than a luxury.

Any technology that can reduce friction and heat inside of a V-8 engine, and by doing so extend the life of the pricey internal parts by minimizing wear, is a good investment. I've heard about oil pump gears being coated with the graphite-based antifriction coatings, and this is another critical point of metal-on-metal contact. Use of friction-reducing coatings can only help here; reduced wear will result in the pump being able to maintain proper internal pump clearances and proper oil pressure for a much longer period of time than a comparable noncoated pump.

Thermal-barrier coatings can be understood as thin layers of porcelain or ceramic. While these coatings can and will heat up, they do a good job of spreading the heat across a wider surface area. This not only minimizes hot spots across the heated surfaces, it also makes it easier for the heat to dissipate. In critical heat-sensitive areas inside V-8 engines, like the tops of pistons, the combustion chamber, the intake and exhaust valves, and the interior surfaces of the intake and exhaust ports, the ability to spread out and dissipate heat is a true performance advantage.

Consider the combustion chamber area. It's not uncommon for temperatures to reach 1,300 degrees Fahrenheit here, and if hot spots develop, preignition and detonation are sure to follow. If the piston tops, valves, and combustion chamber are all coated, detonation is held at bay longer, and more performance can be realized. Being able to run another half-point of compression before encountering detonation is common, and higher cylinder pressures will result in higher power and torque numbers. Also, keeping the incoming air/fuel charge cool and dense has proven to be worth measurable power. By coating the interior of the intake port with a thermal-barrier coating, it serves to "escort" the incoming mix right to the coated intake valve. In a similar vein, coating the exhaust port

These inner and outer springs were treated by Calico Coatings with a graphite-based coating. The friction between inner and outer springs generates intense heat and limits the life of the springs. With the coating, friction and heat are reduced, and the valve spring's life should be extended.

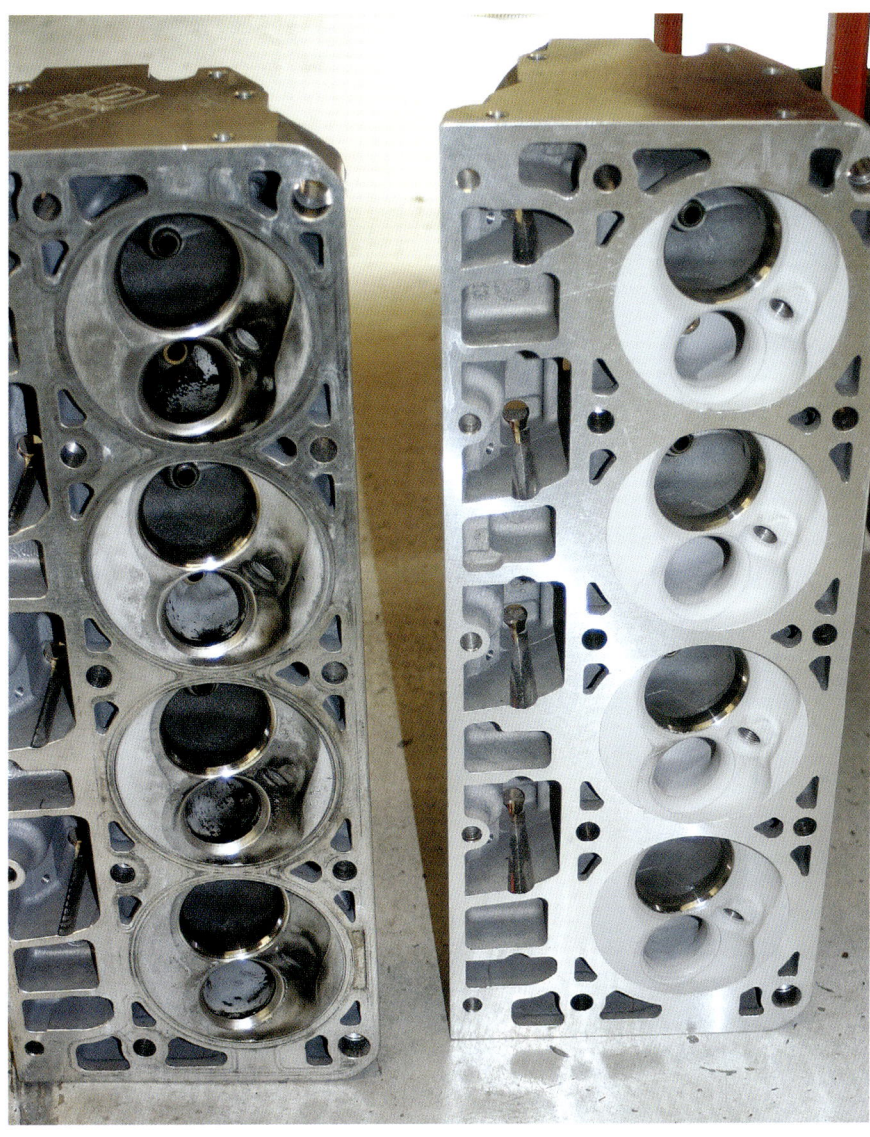

Coating the combustion chambers and ports inside the cylinder head can aid in temperature control, in addition to fighting detonation and preignition. This Pontiac cylinder head has had the ports and chambers coated with a ceramic thermal-barrier coating and the final results are impressive. Keeping the heat in the chamber and spreading it out helps make the chamber more efficient, and porcelain-based coatings like this help make it happen.

helps escort the hot gases past the coated exhaust valve and straight to the header pipe (which can also benefit from coating treatment).

Coatings have also been developed to shed oil. These are particularly of interest when applied to the nonmachined portions of the crankshaft and also to the connecting rods. By shedding unneeded oil from these moving parts, the weight of the oil no longer has to be carried by them. Additionally, this oil will find its way back to the oil pan sooner, which is always good. Oil-shedding coatings become of particular interest when high rpm levels are sustained, as the weight of the oil on the spinning parts becomes more of a concern at higher rpm.

Heat-shedding coatings are also readily available today. These coatings transfer heat more readily and can be a benefit when applied to a traditional V-8 oil pan. Once coated, the entire oil pan becomes a heat sink, and draws unwanted heat

out of the oil. The oil pan becomes an oil cooler this way, and because the addition of the coating is the only modification, there are no issues like fitment or clearance.

If you plan to use coatings ahead of time on a street engine, you could even add some additional compression to wring even more power out of pump gas. Admittedly, coatings are kind of pricey, but I've learned that if you get a lot of parts coated at once, you can usually get a better price. Many performance coatings are designed to be applied at home, but this requires some research before jumping into it. As you'd expect, the surface to be coated must be absolutely free of any dirt or oil so the coating can take hold. Also, different coatings work best when applied at the recommended temperature, so it's crucial the instructions be followed to the letter. A dedicated oven to bake the coatings is also a necessity, and you should never use your home cooking oven for this task.

If you doubt your own ability to apply performance coatings, it's worthwhile to consult a professional. While many professional coating companies exist, I can confidently recommend the internal engine component coating services provided by both TPI Specialties (Chaska, MN, 952-448-6021, tpis.com) and Calico Coatings (Denver, NC, calicocoatings.com). The coatings cannot work for you if they're not applied correctly, and working with an experienced pro will ensure you get the most out of your investment. It's also a good idea to speak with your coating professional of choice about all of your plans and goals ahead of time, as the coater may have some ideas you haven't thought of, or even some new coating products that could serve you well. This area of technology is growing at a rapid pace, and it's wise to see what new coating technologies are out there.

The best thing about internal engine coatings is that another level of engine performance is now available because of their existence. The very best engines still have a bit more performance left in them, and they can be made to last even longer once coatings have been applied. If you are designing and building an engine you plan to rely upon for stout performance over many years, planning for the use of internal engine coatings in strategic locations can improve the mill's power production and long-term durability, and make it better than it could have been without them.

Research into more exotic materials will keep the steady stream of lighter and stronger performance auto parts heading into the aftermarket. Even now, materials like plastic are being used in applications we'd have never thought possible only a decade ago. The best-known of these new material developments includes the plastic intake manifolds found on some new production engines, and subsequent aftermarket versions. The GM LS engine series introduced the feasibility of plastic intake manifolds, and FAST and Weiand have introduced replacement plastic intakes that outperform the OEM piece.

How long will it be before plastic intake manifolds are available for traditional V-8 engines? No one knows, but it seems certain this will happen soon. The LS engine's intake design carries no coolant, which helped allow the use of plastic in its case. But with plastics being lighter than traditional aluminum castings and able to cool so effectively, the potential is there. As plastics research continues, and more exotic materials (and molding techniques) are developed, we can expect to use plastics on more aftermarket components in the future.

One of the composites being developed heavily is carbon fiber. While carbon fiber has been used for exterior body panels on exotic cars for many years, making the jump into the engine bay has been slow. The first engine applications were external, like air filter housings and valve covers, but recently carbon fiber pushrods were developed and offered for sale. While pricey, they were also very lightweight and incredibly strong, and found favor in racing applications, where their cost could be justified.

ALTERNATIVE FUELS

The ever-increasing cost of gasoline has dictated a look toward alternative fuels, including Ethanol/E-85, compressed natural gas (CNG), propane, and even hydrogen generated from fuel cells. Each of these alternatives to gasoline carries with it some genuine benefits as well as a few pitfalls. Should you be considering a switch to an alternative fuel, it would be a great benefit to design and build your V-8 specifically to run on your fuel of choice.

Each of these alternative fuels carries a higher octane rating than traditional gasoline, and if your engine is designed to take advantage of this, improved power production and performance can result. Higher octane means more resistance to detonation under load, which means a higher compression ratio can be tolerated. With octane ratings in the low 100s, compression ratios of 12.5–13:1 are possible, and the resulting power would be comparable to a traditional gasoline engine with the same high-compression ratio.

For ethanol and E-85 users, the proper stoichiometric air/fuel ratio is 10:1, versus 12–14:1 for gasoline. Additionally, the fuel system must be upgraded to deal with the high alcohol content of the fuel, but for many enthusiasts this is simpler than it seems. The original rubber fuel lines and seals used before 1990 or so are not compatible with high alcohol content, which will cause them to break down, become dried out and brittle, and ultimately fail. The newer materials used in modern fuel systems have no such problems. If you've replaced the original seals and rubber in your vintage V-8-powered car, running ethanol or E-85 will not present a problem. The only other issues involve vehicles with plastic fuel tanks. These are also not E-85-compatible. But many later-model vehicles use aluminum or steel tanks, and these are fine with E-85. Vintage domestic vehicles with restored or replacement tanks are also completely compatible with high alcohol content fuels.

Carburetor manufacturers have responded to the growth in E-85 availability with E-85-specific carburetors. Holley, Demon/Barry Grant, and Edelbrock all offer E-85 carburetors with revised jetting and proper seals already installed. This makes such conversions much easier, because the fuel mixer is the primary point of concern, once the rest of the fuel system is up to snuff. Use of E-85 may not even require a radical departure in ignition timing, as the burn rate of E-85 is quite close to that of a traditional gasoline. The only major tuning change involves the 10:1 air/fuel ratio, and this is easy enough to monitor with a commercially available air/fuel gauge with a digital readout. Air/fuel ratio gauges using lights of varying colors are calibrated for use with gasoline, so the digital gauge is a must when tuning on E-85.

The use of composite materials is sure to expand in the future, but we're seeing it begin in earnest now. Shown are an LS-series intake manifold from FAST (top) and the new Pro Flow Lite composite four-barrel carburetor intake from AFR. These lightweight intakes shed heat more readily than the traditional aluminum products they replace. The only downside right now is cost. The tooling and production of these parts still costs much more than traditional aluminum casting and machining. As they become more commonplace, I expect the costs will begin to come down.

What really gets enthusiasts fired up about E-85 is its potential coupled with supercharging. Heat is always a big concern when a supercharger or turbocharger is added. E-85 (or any alcohol-based fuel, like methanol) provides more natural cooling when atomized, making heat less of a worry. Racers using E-85 have reported ice forming on the outside of the supercharger case during drag strip runs under high pressure. Even nonsupercharged E-85-powered performance machine owners have shared stories with me about how cool the intake manifold remains when racing. The potential here is tremendous.

If you're familiar with carburetor rebuilding, you can easily upgrade your existing carb for use with E-85 too. The only major change would be the increased jet sizes the alcohol-based fuels require. While a basic modern rebuild kit will work just fine with E-85, inquiring about an E-85-specific kit isn't a bad idea. Further improvements in seal material technologies could ensure a long and problem-free life for an engine dedicated to life on E-85. Besides the richer fuel delivery requirements and the increased compression ratio, an E-85-specific high-performance V-8 engine build isn't any different than a gasoline-powered one, and all other performance engineering is the same. I'm confident that if E-85 was readily available nationwide, many performance enthusiasts and racers would be using it. The common use of methanol as a performance fuel shows that there's no real secret to it, and the constant pursuit of power justifies such a build. The only holdup has been the infrastructure and distribution of this fuel, as an enthusiast would have to carefully map out any road trip to make sure E-85 was available along the way.

I'm fortunate to live in the Twin Cities of Minnesota, and no other area in the country is as dedicated to E-85 availability. It would be completely feasible to own and drive an E-85-powered hot rod or street machine here, and many do. If the need for E-85 continues to grow, and the availability of new flex-fuel vehicles capable of running on any combination of gasoline and ethanol gains acceptance, this may someday be the case. While many balk at the loss of fuel economy when a vehicle designed for use with gasoline uses E-85, they are hardly taking full advantage of E-85's potential when coupled with forced induction and/or high-compression ratios. For enthusiasts, the performance potential is reason enough to investigate the plausibility of building a high-compression mill to take advantage of what E-85 can deliver.

The availability of CNG and propane is not a question, as both can be had at retailers nationwide. Like E-85, both CNG and propane have higher octane ratings than gasoline, and require greater quantities of fuel when used in automotive applications. The bigger issue with CNG and propane conversions lies in the fuel system upgrades required to support them. Because they must be converted to gaseous form before being introduced to the engine, a relatively complex system of storage (in an approved tank), large-diameter tubing, a cracking device (to heat the liquid fuel and change it to a gas), and a proper mixer (to replace a traditional carburetor) are all required. These components are all readily available from many sources and have been refined for decades, so their engineering and reliability are not in question. What makes these conversions difficult is the adaptation of the necessary components into the project vehicle, and then the need to fill the tank once the vehicle hits the road or track. Liquid gasoline and E-85 are easy enough to transport and dump into the gas tank, but during refueling, CNG and propane must be delivered under high pressure in their liquid state, requiring a trained service technician. If you can locate a retailer willing to service your CNG-propane-powered vehicle, you won't be able to self-serve and you'll be working on their schedule. Filling up will take a bit more time than you're used to, and will require some planning in advance (maybe even an appointment in some cases).

As with E-85, planning for a long-distance road trip in a CNG-propane-powered vehicle will require some research and planning, if it's possible at all. This lack of infrastructure and need for trained service technicians to fill vehicle tanks has held back the idea of CNG-propane-powered cars, but new technologies are emerging to help the cause. There are new self-serve quick-fill pumps on the horizon to ease the pain of filling up, and if these gain popularity and acceptance, things could change very quickly. There are plenty of DOT-approved tanks already in use, as proven in many taxi and postal fleets nationwide. The potential is tremendous if an infrastructure can be built to rival the convenience of gasoline.

In terms of performance, these gaseous fuels hold great potential. There is no concern over gas puddling on the intake manifold floor or fuel coming out of suspension in the incoming airstream like we have with liquid gasoline. The fuel mixers already available have proven to be reliable and could easily be timed to provide high-performance engines with the proper quantities of fuel. Supercharging and turbocharging are common in many industrial CNG/propane engines, and so adding one of these to a CNG- or propane-powered V-8 would be a natural.

One of the newest alternative fuels is hydrogen. When generated from a fuel cell, hydrogen can be burned in a traditional piston engine like any other fuel. While this technology is still on the horizon as of this writing, the potential certainly exists for V-8 engines to be converted to hydrogen use with the addition of a fuel cell. The costs are high right now, as they typically are with any new technology, but in time this may change. The important consideration for fans of V-8 performance is that yet another alternative to traditional gasoline exists and is being heavily researched by automotive manufacturers, and this alone is worthy of mention.

Chapter 13
Supercharging

Designing and building an engine to make good power and last over the long term is a challenging proposition. The addition of a supercharger can make finding big power easier, but it will also require some modifications to improve durability. Many of the modifications we've covered already become essential when a supercharger becomes part of the engine design.

First, we'll discuss the different types of superchargers commonly used on domestic V-8 engines. The most common type of supercharger enthusiasts are familiar with is the Roots design. This is a basic belt-driven air pump that is typically mounted atop the intake manifold. Carburetors or fuel injectors are typically mounted above it. The supercharger compresses the incoming air/fuel mixture before it heads toward the intake valves. There's no question Roots-type

superchargers work well, and they can be tuned to work in both street and competition situations.

A critical part of tuning a Roots blower is the relationship between how fast the supercharger is turned in relation to the crankshaft. This is determined with pulley diameters, and Roots blowers have been run successfully at speeds slower than the crankshaft (underdriven), at the same speed as the crankshaft, and at speeds faster than the crankshaft (overdriven). The key here is in how much air the supercharger can physically move. You might think that the more you spin the blower, the more air moves through it, and therefore there's no potential limit to how much boost you can put into the engine. This is not the case, however. The limiting factors relate to the efficiency of the blower design; above certain rpm, the supercharger is no longer efficient.

For many years, the belt-driven Roots-style supercharger was the standard V-8 supercharger. This one is fed by a pair of vacuum-secondary carburetors, which makes it more street-ready when compared to mechanical secondary carbs. If the engine requires more air and fuel, the carbs will open up, based on a vacuum signal. The vacuum-secondary carbs are identified by the diaphragms on the side of each carburetor body. Roots blowers like this typically require the hood to be modified for clearance, if one is run at all.

There's a lot for engine enthusiasts to see here. While the blown Hemi is enough to get anyone's attention, this hot rod relies on EFI mounted just over the blower, and there's an air filter hiding inside the injector hat. Motorcycle pipe baffles are typically installed in the "zoomie" exhaust headers to quiet things down a little. Little tricks like this can really serve to civilize what looks like a racing engine.

This is the carbureted version of the Magnuson supercharger. The high-tech lobe-coating technology makes it efficient, and those who prefer carbs can enjoy the benefits of this research just like the EFI fans.

Superchargers force additional air (and fuel, if the carb or injectors are mounted before the blower) into an engine to produce more power. This belt-driven blower atop a GM LS-series V-8 make it possible for this engine to make over 500 horsepower using pump gasoline. When teamed with proper EFI computer programming, boosted engines can make big power on demand while remaining mellow enough for street use.

This small belt-driven supercharger is designed for late-model Ford V-8s. Installation is easier than most, because the injectors are already in place, as seen on the side of the intake, alongside the blower case. After installation, a quick software update is all that's needed to get it running right. Things can get more complex when other modifications are made to the engine, like high-lift cams or exhaust headers, that won't perfectly match the computer programming. In cases like this, custom programming of the computer is required, and should be figured into the budget for the project. Custom programming work is expensive, and its cost may come as a surprise if you're not expecting it.

The speed at which a supercharger turns, relative to the engine's crankshaft speed, determines the amount of boost it makes. By changing the diameter of the belt-driven pulley on the supercharger, the amount of boost and at what rpm level it comes in can be altered.

This Vortech centrifugal supercharger is also belt driven, but relies on a spherical compressor design (much the same as the compressor side of a turbocharger) as opposed to the typical Roots-style case design. This makes the centrifugal a different animal. A centrifugal supercharger can be mounted under the hood of most any street machine. They vary greatly in the amount of air they can move, but they've proven to be just as good at making boost as Roots designs. They also are easier to team up with an intercooler, to chill the incoming air before it gets to the engine, and make even more power.

Another look at the centrifugal supercharger shows its drive pulley, which is connected to the engine's crankshaft by a belt. Like the Roots-design supercharger, the diameters of these pulleys can also be altered to fine-tune the boost output to match the needs of the car.

This 454-cubic-inch small-block from Nelson Racing Engines makes 1,150 horsepower and 1,100 ft-lbs of torque on pump gas

The amount of heat generated by the supercharger figures into the limitations of the supercharger. (It is, after all, a compressor, and heat is generated whenever compression occurs.) We know heat can contribute to detonation and, naturally, the overheating of an engine. By compressing the incoming air/fuel mix, we're adding heat before it even enters the chamber. If we add too much heat, detonation is sure to follow, especially if the fuel mixture gets a little lean, which can happen if the supercharger moves more air than the fuel system can support. These issues mean some homework is required to control the boost created by the blower, and coordinate it with the power band of the engine.

Different types of supercharger designs build boost in different ways, and while boost is boost, a Roots blower will act differently than a centrifugal supercharger, which will act differently than a turbocharger.

Centrifugal superchargers are similar to Roots superchargers, in that they are belt driven, but that's about where the similarity ends. They typically build boost more effectively at higher (supercharger) rpm, but by changing pulley diameters, their boost can also be tailored to suit the needs of the engine (and car) they're being added to. Centrifugal blowers are typically mounted away from the intake manifold and rely on tubing to direct the boosted air into the engine. The benefit of this design is that one will comfortably fit under the hood of most domestic vehicles, where most Roots blowers require a hole to be cut into the hood for clearance. (This was more of a concern with carbureted applications mounted atop the blowers than with modern EFI setups, where injectors are under the blower housing.) A wide range of centrifugal blower sizes mean the right size supercharger for your application probably exists, regardless of what that application may be.

Turbochargers don't rely on a belt to turn; they are propelled by exhaust gases. The size of the turbocharger housing is critical, as a smaller turbo will require less exhaust flow to build boost, but will run out of breathing capability sooner than a comparably large turbocharger. The flip side is that a larger turbo will require more exhaust gas flow to even begin building boost, so sizing the turbo to suit your particular needs is critical. This, along with the fact that in a V-8 engine bay, exhaust comes out of opposite sides of the engine, is also why you'll see some

The developments in materials technology have made turbochargers more efficient, and when in used in pairs on a V-8 they can provide incredible results. This twin turbo engine is from Nelson Racing Engines (Chatsworth, California, 818-998-5593, nelsonracingengines.com) and is designed to run on either pump gas (91 octane) or racing gas. The 572-cubic-inch stroker big-block Chevy delivers 1,300 horses and 1,335 ft-lbs, also on pump gas.

If the big-block Chevy isn't intimidating enough, the port-injected nitrous oxide setup ought to be. Nitrous is a great way to add power on demand, as long as the owner doesn't get greedy. Nitrous oxide can be installed and used safely, and can add power to any engine. While this setup is capable of adding several hundred horsepower, a simple plate setup can be dialed in to add 50–150 horsepower at the push of a button.

enthusiasts choosing to run two smaller turbos instead of one large unit. The smaller pair of turbos has the capability to begin building boost sooner, and will provide enough volume of air at higher rpm. The downside is that instead of joining both sides of your exhaust system to feed one turbo effectively, now each side of the engine is tasked with supporting a turbo on its own. The additional expense of a second turbo figures into this choice as well, and research has shown that a single turbo setup is more efficient than running twins.

The upside to a belt-driven supercharger is its relative simplicity. Beyond setting up the belt drive arrangement, there's little to accomplish. Turbos, on the other hand, require relatively extensive plumbing to get the incoming air and exhaust gases to the turbocharger, and then getting air to the engine inlet. The addition of an intercooler (an air-to-air or, in some cases, an air-to-water radiator to cool the incoming compressed air) makes the plumbing challenge even greater, and servicing such an engine regularly can become a real headache. Having so much hot air plumbing under the hood can also wreak havoc on other components, like the braking system or anything made of plastic.

Regardless of the type of supercharging, there are upsides and downsides, but one truth remains constant. Adding boost to your engine will produce more power, and for many hot rodders, that's what it's all about.

There's another way to supercharge your engine we haven't discussed yet: nitrous oxide. The addition of nitrous oxide (and additional fuel) will make more power without question. Nitrous oxide is a chemical consisting of two nitrogen molecules and one oxygen molecule (N_2O). Injecting it into an engine by itself does not add power, but will allow the engine to support the burning of more fuel, due to the presence of the additional oxygen.

The nitrous is stored in a high-pressure bottle (typically between 900 and 1,200 psi) and an electronically activated solenoid allows it to enter the intake manifold under pressure. A second solenoid opens simultaneously to allow more gasoline to be injected at the same time, and the ratio of nitrous to gasoline is controlled by the size of jets installed in the respective injectors.

The overall system design is rather simple, and nitrous is typically introduced to the intake though a plate installed

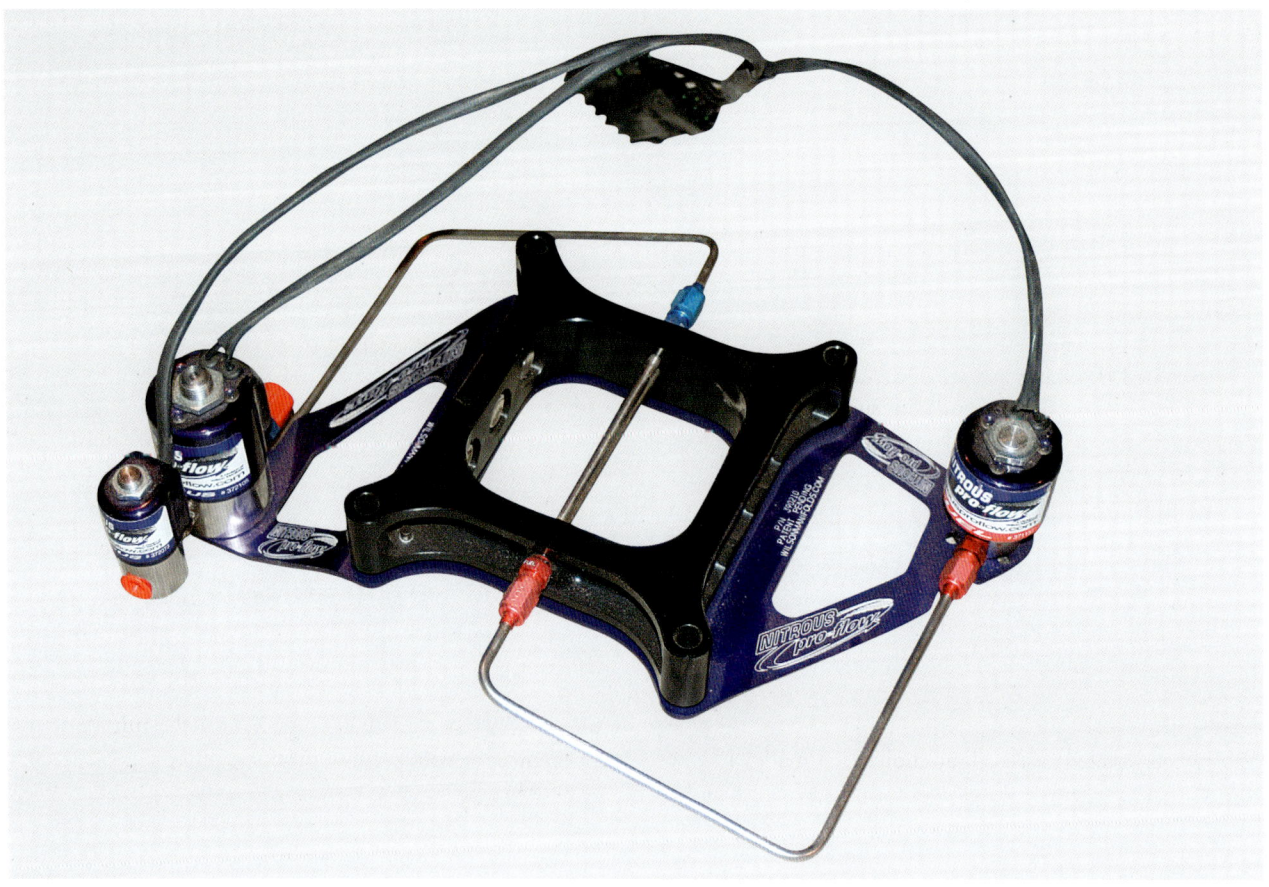

Basic nitrous oxide setups typically use plates like this one. The gaseous nitrous oxide and additional fuel are added when the solenoids mounted on the side of the plate are energized and open. Jets in the nitrous and fuel feed lines can be changed to alter the fuel-to-nitrous ratio, or to add more power (if larger jets are installed). This setup, from Nitrous Pro-Flow, is well engineered for reliable performance.

under the carburetor or directly into each individual intake port. The difficulty arises when you decide how much nitrous to add, and for how long.

Because nitrous oxide injection is a "power on demand" type of supercharger (as opposed to belt-driven superchargers and turbos, which add power all the time) you have the power to decide when it's required. You also have a limited amount of power; the fun ends when the bottle is empty. If you feel a limited amount of additional power-on-demand is what you're after, nitrous is tough to beat.

Many fear the power of nitrous, as greedy hot rodders have hurt many an engine while using it. The key is not to use too much for too long. Nitrous can be a safe and sane way to get additional power only when you need it.

All types of superchargers will use more fuel in order to make more power. For this reason, an upgraded fuel system is an absolute requirement when adding a supercharger to any engine. The only possible exception might be a very small (50–100 horsepower) nitrous setup, but even then I'd still upgrade the fuel pump and add a pressure regulator to ensure consistent supply at the correct pressure. Like I mentioned earlier, if a supercharged system of any kind goes lean, damage will result.

Having a capable and reliable source of adequate fuel is essential with any type of supercharger.

I must encourage you to research your supercharger of choice thoroughly, and once you've decided on a supercharger type, work with a single manufacturer to get it all dialed in. The makers of belt-driven superchargers, turbochargers, and nitrous components typically offer complete kits designed to work as a system, and I've found these are the best ways to go.

Like my recommendations for camshaft, valvetrain, and fuel systems, I've learned that teaming components from the same manufacturer ensures that everything will work together the way it's supposed to, and this is especially critical if you've never done something like this before. One-stop shopping in this manner assures you a single source to answer any questions that may arise, and if you're new to the art of supercharging, that is a critical resource. If you're trying to make a series of components work together that were never designed to operate in sync, you might run into problems only experience can answer. Sometimes those lessons can be very expensive, so until you've gained enough experience to call yourself an expert, work with those who have proven they and their products can and will work for you.

Chapter 14
Summary

THE BIG PICTURE

When developing a V-8 engine for increased performance, defining the true purpose of your engine is absolutely critical, and this must be done long before any components are purchased. Looking at each individual system in light of your goals will surely result in success. Once you've determined where you want your engine's powerband to be, where your upper rpm limit will be, and how long the engine will be designed to last, then you can consider the budget you've got to work with.

Determining the best possible displacement, compression ratio, camshaft, intake system, exhaust system, oiling system, and cooling system for your specific purpose will guide you to the right choices. Even those working with a tight budget have options. Even combining the proper factory parts can result in better performance than the factory ever offered, if the right choices are made. The key is to design your engine from top to bottom as a matched

series of systems to deliver the best possible performance with the least possible compromise.

I've found it never hurts to get advice, but you have to remember who you're getting advice from. I like to consult with several experts who have different backgrounds, then compare their recommendations with my own conclusions. It's not unusual for a cam manufacturer to suggest one cam, a race engine builder to suggest another, and a street performance enthusiast to suggest a third for the exact same engine. This is why it pays to know what you're talking about and to ask the right questions. Once you've gotten plenty of advice, you can make your own choices, and the odds are strong you'll be pleased with the final results.

One last thing to consider when designing and building the engine of your dreams is the external equipment and appearance it will present. Making sure a good-quality series of belt-driven accessories can be engineered into place

Builders often overlook the necessary accessories bolted to the front of a performance engine. Then the owner is stuck trying to figure it all out. Vintage-style V-belts are fine in nostalgic or low-performance setups, but if you're getting serious about high-rpm performance, it's time to consider more modern ribbed belts, serpentine belts, or even cog (toothed) belts. This setup is from Jones Racing Components, and is a great example of a well-engineered aftermarket drive system. It uses a serpentine belt from the crank to drive the water pump and power steering pump, while a cog belt is used (from the water pump) for the alternator.

Ignition timing is a basic tuning element with any engine, and is particularly important with a high-performance build. Using a top-quality timing light (like this one from Snap-on) is the only correct way to check and adjust your engine's timing. Some of the features in this tool include the ability to advance and retard the strobe (so you can always use the TDC timing mark, which is easiest to see) and a tachometer (so you can monitor engine rpm while checking timing).

Another necessary tool is a good-quality torque wrench, like this one. Once you have a good torque wrench, it's wise to have it calibrated for accuracy every year. So many automotive engine fasteners require specific torque values, it's critically important to have a quality, calibrated torque wrench on hand.

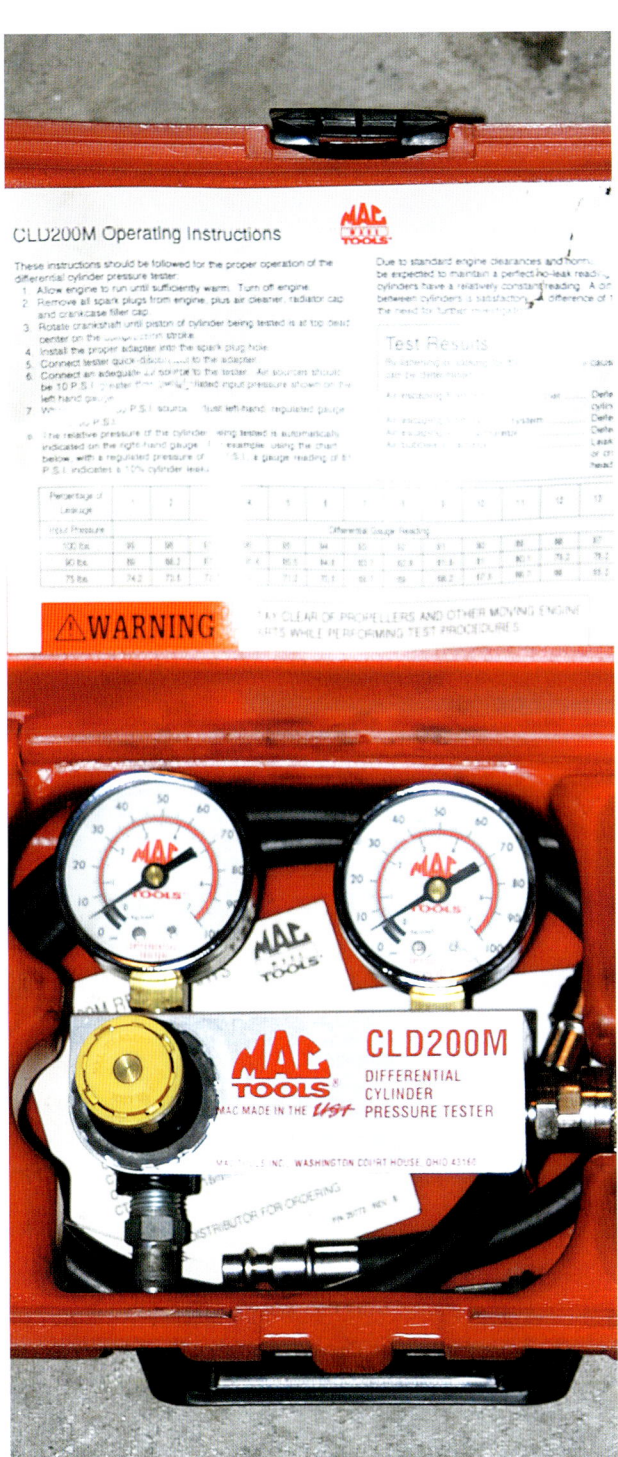

One tool more enthusiasts should own is a cylinder leakdown tester. This one, from Mac Tools, is typical. It directs compressed air into a cylinder (at TDC on the compression stroke) through the spark plug hole. The gauges compare the pressure of the air being fed into the cylinder with the air inside it, and the difference reveals the amount of air leaking out. This test is done with the engine at operating temperature. If more than 7 psi is leaking from any one cylinder, it would be wise to investigate the most probable points for the cylinder to leak: the piston rings and the valves. Also, by testing each cylinder, if one is much worse than all of the others, you can focus your investigation there. All the cylinders readings should be within a couple of psi of each other.

Maintaining a high-performance engine is the key to helping it live a long life. With a solid (not hydraulic) flat tappet or roller camshaft, checking valve lash is a regular part of the maintenance program. Get a good set of feeler gauges and check lash a couple times a year. The correct lash settings are called out by the manufacturer of your camshaft on the cam card. They should be checked while the engine is at normal operating temperature for the highest degree of accuracy.

After investing heavily in a performance engine, get a good set of gauges to keep an eye on its functions. Temperature, oil pressure, and electrical system function (either amps or volts) represent the basic needs. An oil temperature gauge and an exhaust-mounted oxygen sensor would also be wise choices.

is critical, and a little forethought here can save headaches down the road.

You also must have the proper tools on hand to maintain this dream engine of yours. I'm not only talking about the special tools some makes of engine or specialized multicarb setups require to be tuned properly. You need high-quality tune-up and assembly tools too. Trying to check timing on a big-dollar engine with a cheap timing light is asking for trouble. If you've not cut any corners on quality throughout your engine build, why start cutting corners with its maintenance tools? It simply does not make sense.

In a similar common sense vein, invest in some good-quality gauges to monitor the health of your new engine. Keeping track of the powerplant's status is as easy as installing some basic sensors and wiring. The ability to monitor the status of the engine at any time can alert you to potential issues before they cause expensive damage, and you'd be wise to spot problems early, rather than trying to figure out what went wrong afterward.

The appearance of your engine is something many plan for from the start, and while many vintage and custom speed parts do not work as well as their modern counterparts, this does not mean they cannot be made to work quite well. Learning to work with unusual equipment can be frustrating sometimes, but the lessons learned and the problems solved make you a better technician and allow you to have experience—and an underhood look—to which few others can lay claim.

Hopefully, the lessons and suggestions offered here will equip you to make the best possible choices, and the engines you'll be building in the future will deliver even more performance than you expected. A truly great engine will be more than powerful; it will also be durable and reliable. The great American V-8 engines are fully capable of being all these things, and with some research and effort, you'll be able to enjoy the full capabilities of your own custom-designed and custom-built engine. I hope this book assists you, and I encourage you to share the book and the information it contains with the next generation of engine builders. This will ensure that they will be able to experience the same thrill of V-8 power that we do, and the same satisfaction of knowing they did it themselves.

SUMMARY

Just how much impact do some vintage speed parts make? Without the early tri-power setup, how exciting would this early powerplant be?

This Windsor-based Ford small-block V-8 could have been outfitted with EFI or a new-style supercharger, but it just looks better with triple carbs when it's under the hood of a traditional hot rod. Choosing the right parts to nail the look you're after is almost as important as making them work to the best of their ability.

Index

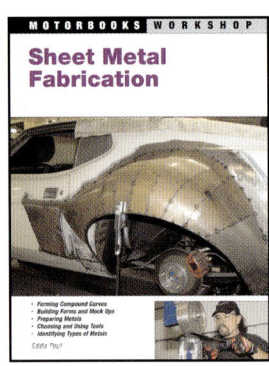

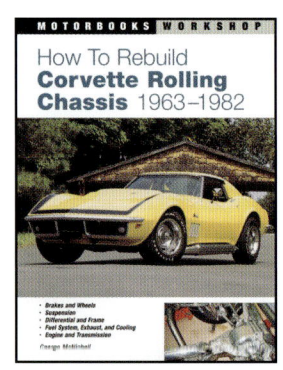

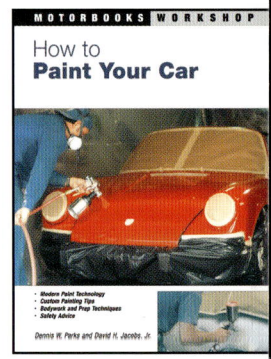

Other Great Books in this Series

How to Paint Your Car
136261AP • 978-0-7603-1583-5

How to Paint Flames
137414AP • 978-0-7603-1824-9

How to Master Airbrush
Painting Techniques
140458AP • 978-0-7603-2399-1

How to Repair Your Car
139920AP • 978-0-7603-2273-4

How to Diagnose
and Repair Automotive
Electrical Systems
138716AP • 978-0-7603-2099-0

Chevrolet Small-Block
V-8 ID Guide
122728AP • 978-0-7603-0175-3

How to Restore and
Customize Auto
Upholstery and Interiors
138661AP • 978-0-7603-2043-3

Sheet Metal
Fabrication
144207AP • 978-0-7603-2794-4

101 Performance Projects For Your
BMW 3 Series 1982–2000
143386AP • 978-0-7603-2695-4

Honda CRF Performance Handbook
140448AP • 978-0-7603-2409-7

Autocross Performance Handbook
144201AP • 978-0-7603-2788-3

Mazda Miata MX-5
Find It. Fix It. Trick It.
144205AP • 978-0-7603-2792-0

Four-Wheeler's Bible
135120AP • 978-0-7603-1056-4

How to Build a Hot Rod
135773AP • 978-0-7603-1304-6

How to Restore Your Collector Car
128080AP • 978-0-7603-0592-8

101 Projects for Your
Corvette 1984–1996
136314AP • 978-0-7603-1461-6

How to Rebuild Corvette Rolling
Chassis 1963–1982
144467AP • 978-0-7603-3014-2

How to Restore Your Motorcycle
130002AP • 978-0-7603-0681-9

101 Sportbike
Performance Projects
135742AP • 978-0-7603-1331-2

How to Restore and Maintain Your
Vespa Motorscooter
128936AP • 978-0-7603-0623-9

How to Build a Pro Streetbike
140440AP • 978-0-7603-2450-9

101 Harley-Davidson Evolution
Performance Projects
139849AP • 978-0-7603-2085-3

101 Harley-Davidson Twin Cam
Performance Projects
136265AP • 978-0-7603-1639-9

Harley-Davidson Sportster
Performance Handbook,
3rd Edition
140293AP • 978-0-7603-2353-3

Motorcycle Electrical Systems
Troubleshooting and Repair
144121AP • 978-0-7603-2716-6